Lecture Notes in Artificial Intelligence

Edited by R. Goebel, J. Siekmann, and W. Wahlster

Subseries of Lecture Notes in Computer Science

David Riaño   Annette ten Teije
Silvia Miksch   Mor Peleg (Eds.)

Knowledge Representation for Health-Care

Data, Processes and Guidelines

AIME 2009 Workshop KR4HC 2009
Verona, Italy, July 19, 2009
Revised Selected and Invited Papers

 Springer

Series Editors

Randy Goebel, University of Alberta, Edmonton, Canada
Jörg Siekmann, University of Saarland, Saarbrücken, Germany
Wolfgang Wahlster, DFKI and University of Saarland, Saarbrücken, Germany

Volume Editors

David Riaño
Universitat Rovira i Virgili
Av. Països Catalans 26, 43007 Tarragona, Spain
E-mail: david.riano@urv.net

Annette ten Teije
Vrije Universiteit Amsterdam
De Boelelaan 1081A, 1081HV Amsterdam, The Netherlands
E-mail: annette@cs.vu.nl

Silvia Miksch
Danube University Krems
Dr.-Karl-Dorrek-Str. 30, 3500 Krems, Austria
E-mail: silvia.miksch@donau-uni.ac.at

Mor Peleg
University of Haifa
Department of Management Information Systems
Rabin Bldg., 31905 Haifa, Israel
E-mail: morpeleg@mis.hevra.haifa.ac.il

Library of Congress Control Number: 2010920464

CR Subject Classification (1998): H.3, H.5.2, J.3, H.2.8, I.2.6, E.1

LNCS Sublibrary: SL 7 – Artificial Intelligence

ISSN   0302-9743
ISBN-10   3-642-11807-0 Springer Berlin Heidelberg New York
ISBN-13   978-3-642-11807-4 Springer Berlin Heidelberg New York

springer.com

© Springer-Verlag Berlin Heidelberg 2010

Typesetting: Camera-ready by author, data conversion by Scientific Publishing Services, Chennai, India
Printed on acid-free paper   SPIN: 12984831   06/3180   5 4 3 2 1 0

Preface

This book is the result of merging two workshops series, namely, one on computerized guidelines and protocols and the other one on knowledge management for health care procedures. The merge resulted in the KR4HC workshop: Knowledge Representation for Health Care: Data, Processes, and Guidelines. This workshop was held in conjunction with the 12th Conference on Artificial Intelligence in Medicine (AIME 2009), in Verona, Italy. The book included, in addition to the full-length workshop papers, invited peer-reviewed advanced papers on lessons learned in these fields.

The KR4HC workshop continued a line of successful guideline workshops held in 2000, 2004, 2006, 2007, and 2008. Following the success of the first European Workshop on Computerized Guidelines and Protocols held in Leipzig, Germany, in 2000, the Symposium on Computerized Guidelines and Protocols (CGP 2004) was organized in Prague, Czech Republic in 2004 to identify use cases for guideline-based applications in health care, computerized methods for supporting the guideline development process, and pressing issues and promising approaches for developing usable and maintainable vehicles for guideline delivery. In 2006 an ECAI 2006 workshop at Riva del Garda, Italy, entitled "AI Techniques in Health Care: Evidence-Based Guidelines and Protocols" was organized to bring together researchers from different branches of artificial intelligence to examine cutting-edge approaches to guideline modeling and development and to consider how different communities can cooperate to address the challenges of computer-based guideline development. This ECAI 2006 workshop continued with a workshop on "Computer-Based Clinical Guidelines and Protocols (CCG 2008)" held at the Lorentz Centre of Leiden University at the beginning of 2008, which resulted in the book "Computer-Based Clinical Guidelines and Protocols: A Primer and Current Trends" edited by Annette ten Teije, Silvia Miksch, and Peter Lucas and published by IOS Press in 2008.

Running in parallel to the previous workshops, the KR4HC workshop was the sixth in a series of workshops and publications devoted to the formalization, organization, and deployment of procedural knowledge in health care. These previous workshops and publications are the IEEE CBMS 2007 special track on "Machine Learning and Management of Health Care Procedural Knowledge" held in Maribor, Slovenia in 2007; the AIME 2007 workshop entitled "From Medical Knowledge to Global Health Care" held in Amsterdam, The Netherlands, in 2007; the ECAI 2008 workshop on "Knowledge Management for Health Care Procedures" in Patras, Greece, in 2008, and the Springer *Lecture Notes in Artificial Intelligence* books LNAI 4924 and LNAI 5626, both edited by David Riaño in 2008 and 2009, respectively.

As computerized health care support systems are rapidly becoming more knowledge intensive, the representation of medical knowledge in a form that enables reasoning is growing in relevance and taking a more central role in the area of medical informatics. In order to achieve a successful decision-support and knowledge management approach to medical knowledge representation, the scientific community has to provide efficient representations, technologies, and tools to integrate all the important elements that health care providers work with: electronic health records and health care information systems, clinical practice guidelines and standardized medical technologies, codification standards, etc.

Synergies to integrate the above-mentioned elements and types of knowledge must be sought both in the medical problems (e.g., prevention, diagnosis, therapy, prognosis, etc.) and also in the computer science and artificial intelligence technologies (e.g., natural language processing, digital libraries, knowledge representation, knowledge integration and merging, decision support systems, machine learning, e-learning, etc.).

This book presents 11 selected and extended papers out of 23 submissions of the Workshop on "Knowledge Representation for Health Care: Data, Processes and Guidelines (KR4HC 2009)". All extended papers were reviewed by at least two reviewers and revised accordingly. The topics range from patient data management, maintaining and extracting medical ontologies, temporal representation and reasoning to guideline and protocol design, execution, and dissemination as well as the integration of electronic patient records into guideline-based care and decision-support systems.

We invited four well-known researchers in the scientific community to submit state-of-the-art papers, which were reviewed by at least two reviewers and revised accordingly. Dionisio Acosta, Vivek Patkar, Mo Keshtgar, and John Fox propose a computational framework to provide a clinical guideline-based decision support system for breast cancer multidisciplinary meeting. Silvia Panzarasa, Silvana Quaglini, Anna Cavallini, Giuseppe Micieli, Simona Marcheselli, and Mario Stefanelli address the challenging topic of integrating a decision model used to represent and execute guideline recommendations with end-user interface and the electronic patient record. Mor Peleg presents an approach to sharing computer-interpretable guidelines with more than one implementing institution. David Riaño introduces a knowledge management architecture to integrate, maintain, and share medical and clinical data, information and knowledge.

Thanks should go to the people who contributed to the KR4HC 2009 workshop: the authors of the submitted papers, the authors of the invited papers, the members of the Organizing Committee, the members of the Program Committee and the sponsor institutions.

We aim to organize KR4HC every year in conjunction with a Medical Informatics or Artificial Intelligence conference in order to offer a stable platform for the interaction of the community in the area of knowledge representation for health care.

December 2009

David Riaño
Annette ten Teije
Silvia Miksch
Mor Peleg

Organization

The workshop "Knowledge Representation for Health-Care: Data, Processes, and Guidelines" and the edition of this book were organized by D. Riaño (Rovira i Virgili University, Tarragona, Spain), A. ten Teije (Vrije Universiteit Amsterdam, Amsterdam, The Netherlands), S. Miksch (Danube University Krems, Krems, Austria), and M. Peleg (University of Haifa, Haifa, Israel).

Program Committee

Syed Sibte Raza Abidi	Dalhousie University, Canada
Ameen Abu-Hanna	University of Amsterdam, The Netherlands
Roberta Annicchiarico	Santa Lucia Hospital, Italy
Luca Anselma	Università di Torino, Italy
Fabio Campana	CAD RMB, Italy
Paul de Clercq	University of Maastricht, The Netherlands
John Fox	University of Oxford, UK
Robert Greenes	Harvard University, USA
Femida Gwadry-Sridhar	University of Western Ontario, Canada
Frank van Harmelen	Vrije Universiteit Amsterdam, The Netherlands
Tamás Hauer	CERN, Switzerland
Jim Hunter	University of Aberdeen, UK
Katharina Kaiser	Vienna University of Technology, Austria
Patty Kostkova	City University London, UK
Peter Lucas	University Nijmegen, The Netherlands
Mar Marcos	Universitat Jaume I, Castellon, Spain
Silvia Miksch	Danube University Krems, Austria
Stefani Montani	Università del Piemonte Orientale, Alessandria, Italy
Mor Peleg	University of Haifa, Israel
Silvana Quaglini	University of Pavia, Italy
David Riaño	Rovira i Virgili University, Spain
Kitty Rosenbrand	Dutch Institute for Healthcare Improvement (CBO), The Netherlands
Yuval Shahar	Ben Gurion University, Beer-Sheva, Israel
Brigitte Seroussi	STIM, DPA/DSI/AP-HP, France
Andreas Seyfang	Vienna University of Technology, Austria
Robert Stevens	University of Manchester, UK
Maria Taboada	University of Santiago de Compostela, Spain
Annette ten Teije	Vrije Universiteit Amsterdam, The Netherlands

Paolo Terenziani   Università del Piemonte Orientale Amedeo
                           Avogadro, Italy
Samson Tu   Stanford University, USA
Aida Valls   Rovira i Virgili University, Spain
Dongwen Wang   University of Rochester, USA
Jeremy Wyatt   National Institute of Clinical Excellence, UK

Table of Contents

From Patient Data to Medical Ontologies

Creating Topic Hierarchies for Large Medical Libraries.............. 1
David Sánchez and Antonio Moreno

Bridging an Asbru Protocol to an Existing Electronic Patient Record ... 14
Claudio Eccher, Andreas Seyfang, Antonella Ferro,
Sergey Stankevich, and Silvia Miksch

From Natural Language Descriptions in Clinical Guidelines to
Relationships in an Ontology 26
María Taboada, María Meizoso, David Riaño, Albert Alonso, and
Diego Martínez

A Hybrid Methodology for Consumer-Oriented Healthcare Knowledge
Acquisition.. 38
Elena Cardillo, Luciano Serafini, and Andrei Tamilin

Identifying Disease-Centric Subdomains in Very Large Medical
Ontologies: A Case-Study on Breast Cancer Concepts in SNOMED
CT. Or: Finding 2500 Out of 300.000 50
Krystyna Milian, Zharko Aleksovski, Richard Vdovjak,
Annette ten Teije, and Frank van Harmelen

Sharable Appropriateness Criteria in GLIF3 Using Standards and the
Knowledge-Data Ontology Mapper 64
Mor Peleg

Guideline Modeling and Tools

Analysis of the GLARE and GPROVE Approaches to Clinical
Guidelines 76
Alessio Bottrighi, Federico Chesani, Paola Mello, Marco Montali,
Stefania Montani, Sergio Storari, and Paolo Terenziani

Semantic Web-Based Modeling of Clinical Pathways Using the UML
Activity Diagrams and OWL-S 88
Ali Daniyal and Syed Sibte Raza Abidi

Extracting Qualitative Knowledge from Medical Guidelines for Clinical
Decision-Support Systems.. 100
Maarten van der Heijden and Peter J.F. Lucas

Experiences in the Development of Electronic Care Plans for the
Management of Comorbidities 113
    Esther Lozano, Mar Marcos, Begoña Martínez-Salvador,
    Albert Alonso, and Josep Ramon Alonso

Challenges in Delivering Decision Support Systems: The MATE
Experience 124
    Dionisio Acosta, Vivek Patkar, Mo Keshtgar, and John Fox

Technical Solutions for Integrating Clinical Practice Guidelines with
Electronic Patient Records 141
    Silvia Panzarasa, Silvana Quaglini, Anna Cavallini,
    Giuseppe Micieli, Simona Marcheselli, and Mario Stefanelli

Advanced Topics

Towards a Possibility-Theoretic Approach to Uncertainty in Medical
Data Interpretation for Text Generation 155
    François Portet and Albert Gatt

Argumentation about Treatment Efficacy 169
    Nikos Gorogiannis, Anthony Hunter, Vivek Patkar, and
    Matthew Williams

A Knowledge-Management Architecture to Integrate and to Share
Medical and Clinical Data, Information, and Knowledge 180
    David Riaño

Author Index 195

Creating Topic Hierarchies for Large Medical Libraries

David Sánchez and Antonio Moreno

ITAKA-Intelligent Technologies for Advanced Knowledge Acquisition
Department of Computer Science and Mathematics,
University Rovira i Virgili
Av. Països Catalans, 26. 43007, Tarragona, Spain
{david.sanchez,antonio.moreno}@urv.cat

Abstract. Web-based medical digital libraries contain a huge amount of valuable, up-to-date health care information. However, their size, their keyword-based access methods and their lack of semantic structure make it difficult to find the desired information. In this paper we present an automatic, unsupervised and domain-independent approach for structuring the resources available in an electronic repository. The system automatically detects and extracts the main topics related to a given domain, building a taxonomical structure. Our Web-based system is integrated smoothly with the digital library's search engine, offering a tool for accessing the library's resources by hierarchically browsing domain topics in a comprehensive and natural way. The system has been tested over the well-known PubMed medical library, obtaining better topic hierarchies than those generated by widely-used taxonomic search engines employing clustering techniques.

Keywords: taxonomy learning; digital libraries; Web mining; Web search engines; topic hierarchies; knowledge acquisition; ontologies; Semantic Web.

1 Introduction

The last century has brought huge advances in the medical field. Nowadays there are hundreds of scientific journals that provide the latest evidence-based data. Medical practitioners need to be aware of all this information in a timely basis, to provide the best possible care in all the phases of diagnosis, treatment and prognosis. With the growth of the Information Society, new ways of accessing this enormous amount of medical information have emerged. Classical physical supports or local databases have been replaced by electronic repositories and Web portals. Medical Web-based digital libraries (e.g. PubMed[1], EMBASE[2], OVID[3], MedLine[4], UMLS[5], OpenMed[6],

---

[1] http://www.ncbi.nlm.nih.gov/pubmed/
[2] http://www.embase.com/
[3] http://gateway.ovid.com/
[4] http://medlineplus.gov/
[5] http://www.nlm.nih.gov/research/umls/
[6] http://openmed.nic.in/

D. Riaño et al. (Eds.): KR4HC 2009, LNAI 5943, pp. 1–13, 2010.

UK's National electronic libraries[7], US' National Library of Medicine[8], etc.) represent valuable sources in which the medical knowledge and scientific results are stored, providing a trusted, updated and accessible environment for professionals and researchers [19].

However, the success of these initiatives has resulted in continuously growing repositories in which the amount of resources is so huge that the difficulty of searching and obtaining the desired information has become a serious problem [21]. Due to this fact, there is a need for automated tools for information indexing and retrieval that ease the way in which those resources are searched and analyzed.

The usual way to access these resources is to use the keyword-based search engines provided by the digital libraries (PubMed is the case study in this paper). They are useful for retrieving relevant resources for a given query, but this type of search usually suffers from two problems derived from the nature of textual queries and the lack of semantic analysis: *a)* the difficulty to set the most appropriate and restrictive query for the searched information, and *b)* the tedious evaluation of the potentially huge amount of obtained resources.

In order to minimize those problems, PubMed incorporates a certain degree of structure (e.g. resources can be queried by authors, title, references, etc.) and supports predefined medical ontologies and thesaurus such as MeSH[9] or EMTREE[10] in order to ease the query formulation. However, it lacks a *semantic* structure constructed according to the topics which are actually covered by the stored documents. In a narrower context those structures may be managed by hand but, in the case of wide repositories such as PubMed, the amount of resources to deal with is so enormous[11] that it makes unfeasible to manually construct or maintain topic classifications.

In order to tackle this problem, this paper presents and describes a tool to automatically generate topic hierarchies according to the information stored in the resources of a digital library. It relies on syntactical and semantic analysis of natural language resources of a given domain. Being unsupervisedly created, those hierarchies can be used to easily construct up-to-date semantic indexes of the library's resources according to their actual content. Although the approach is domain-independent (as no predefined knowledge is needed, avoiding to introduce limitations in its applicability), it is particularly useful in very dynamic and knowledge-intensive areas like Medicine. In particular, we have applied and evaluated it on PubMed.

The rest of the paper is organised as follows. Section 2 presents related works on the automatic construction of taxonomies and topic hierarchies from digital repositories. Section 3 details the proposed methodology to obtain topic hierarchies from digital libraries. Section 4 explains how the hierarchies can be used to help the user to browse the library's resources. Section 5 evaluates the results obtained in PubMed for several medical domains, comparing them to a widely-used cluster-based search engine. The final section contains the conclusions and proposes some lines of future work.

---

[7] http://www.library.nhs.uk/Default.aspx

[8] http://www.nlm.nih.gov/

[9] http://www.nlm.nih.gov/mesh/

[10] http://www.ovid.com/site/products/fieldguide/embx/EMTREE_Thesaurus.jsp

[11]PubMed has more than 18 million citations (Jan09, http://en.wikipedia.org/wiki/PubMed).

2 Related Work

Nowadays, there exist several initiatives for structuring electronic repositories and, more generally, the whole Web in an unsupervised way. However, most of the approaches are based on clustering techniques: they group similar objects into sets. In the Web search context, this implies the organization of Web pages into groups, so that different groups correspond to different user needs. Some systems of this kind are: Scatter/Gather [9], which provides online -fast- and offline -slow- clustering algorithms; Grouper [30] which operates on query result snippets and clusters together documents with large common subphrases; Clusty (http://clusty.com), for commercial purposes; Mapuccino [23], using similarity based on the vector-space model; and SHOC [31] based on key phrase discovery.

The main problem of those approaches is the poor semantics of the structure presented to the user: clusters obtained for a searched domain tend to be unconnected, with very different degrees of granularity. This hampers the comprehension of the domain structure and the browsing of the available resources.

On the other hand, *taxonomies* are a typical way of structuring knowledge. In Artificial Intelligence systems, in fact, they are the first step for composing *ontologies* [13] and a crucial component in many knowledge-intensive environments like the Semantic Web [1]. Nowadays, there exist *taxonomy generator* packages, such as the Autonomy/Verity Thematic Mappings [5], which allow the construction of topic hierarchies that can be used for browsing or classification. Taxonomical approaches have been applied for structuring Web resources in *directory services* (like Yahoo), offering a very intuitive way of browsing. However, they are generally constructed manually by subject matter experts [14]. This requires a considerable amount of human effort, and it is hardly scalable in an enormous and dynamic repository like the Web. In consequence, directory services have a limited scope, in comparison to a general Web search engine, and are incomplete, especially in technological domains. The same situation, albeit in a smaller scale (millions instead of billions of resources), applies to digital repositories.

Due to the described limitations of clustering-based methods and of the manual construction of taxonomies, we aim to design an automatic and unsupervised tool for taxonomically structuring the resources of an electronic library.

Our approach is based on knowledge acquisition from text. The most well-known approaches in this area are: *pattern-based extraction* [18, 24] where a relation is recognized when a sequence of words in the text matches a pattern; *association rules* [2]; *ontology pruning* [20] that is based on refining a general ontology using heterogeneous sources, and *concept learning* [17] where a given taxonomy is incrementally updated as new concepts are acquired from texts. However, most of these approaches have been applied for developing methodologies that deal with a pre-selected domain corpus and/or use some degree of supervision and previous knowledge [15]. Those aspects hamper their performance when dealing with a huge and unstructured repository.

Recently, some authors have been using the Web as a learning corpus for developing [25] or enriching knowledge structures [1], proposing techniques adapted to this particular environment [27]. In particular, statistical analysis have been applied for ranking synonym sets [29] or checking the relevance of pattern-based extracted candidates for taxonomic relationships [7]. These statistics, in conjunction with an exhaustive utilization of Web search engines, have been used for acquiring large lists of facts using a set of predefined linguistic patterns [11]. Those are some of the basis of our approach which, as will be shown in the evaluation section, is able to improve the results provided by some of the introduced approaches.

3   Automatic Learning of Topic Hierarchies

The base of our proposal is the semantic analysis of the documents available in the electronic repository in order to extract topic candidates. Those documents are retrieved by means of the library's search engine. A topic's relevance for the input domain is assessed by means of a Web-scale statistical analysis based on the Web's information distribution.

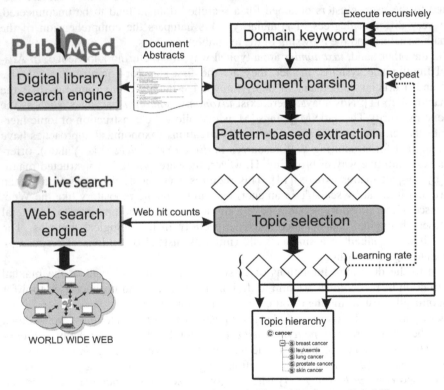

Fig. 1. General schema for constructing topic hierarchies from a digital library

As shown in Figure 1, the algorithm starts from a keyword indicating the domain for which the topic hierarchy should be created (e.g. *cancer*). Then, it executes three tasks: *i)* retrieval of the documents of the library related to the given domain, *ii)* extraction of candidates that may represent different subtopics for the domain via semantic analysis of the content of the documents and *iii)* selection of the most reliable candidates based on a Web-scale statistical analysis. These steps are explained in the following sections.

3.1 Document Retrieval

In order to retrieve the domain-related documents from which to discover relevant topics, the system uses the initial keyword to query the library's search engine. It is mandatory that the digital library incorporates a keyword-based search engine (like PubMed does). The text contained in the obtained documents is syntactically analysed and scanned to discover semantic regularities which may express specializations (subtopics) of the main concept.

However, as the list of resources to be analysed can be large (e.g. near 150,000 documents are returned for the *anaemia* domain in the PubMed repository), this approach may not be scalable enough. For this reason, as shown in Figure 1, we introduce two strategies to improve the performance of document analysis:

- In many cases, it is possible to obtain previews, abstracts or summaries. Considering that summaries obtained by a keyword-matching search engine contain the main topic of a resource, it is usually possible to extract relevant semantic information from them. In this manner, we can reduce the dimensionality of the search both in the number of queries (one query may return dozens of summaries) and the size of the text to be analysed (an abstract represents a minimal part of a document's text).
- Even considering the above strategy, there can be thousands of result pages that should be accessed and analysed. However, relevant knowledge is typically redundant [8] and, in consequence, it can be potentially retrieved from a reduced set of documents. For that reason, in our approach, only a reduced number of summaries is analyzed. As will be described in section 3.4, the system automatically decides the number of resources to analyse according to the domain's generality and the potential amount of available subtopics.

3.2 Candidate Topic Extraction

As stated in section 2, the construction of a topic hierarchy for a given domain based on textual resources is a problem closely related to the one of taxonomy learning from texts. Considering that taxonomies represent semantic relationships between concepts, a common learning methodology is the use of *linguistic patterns*.

Pattern-based approaches are heuristic methods in which the text is scanned for instances of distinguished lexical-syntactic patterns that indicate a relation of interest. They are especially useful for detecting specialisations of concepts that can represent *is-a* (taxonomic) relations [18]. Several authors such as Grefenstette [16] and Hearst [18] have studied and defined those patterns for obtaining taxonomical relationships from English-written texts (see Table 1). These patterns are applied over the text retrieved in the previous stage to detect *candidate concepts* which may represent subtopics of the given domain.

In more detail, the system starts by analyzing the first set of abstracts (e.g. 20) presented by the library's search engine for the queried domain. The content is processed in order to extract clear text and, after syntactically analysing it, it is parsed in order to discover pattern matchings. Each matching is analysed in order to extract potential hyponyms.

Table 1. Linguistic patterns expressing taxonomic relationships (NP=Noun Phrase)

Pattern	Example	Relation
{ADJ|NOUN}$^+$ NOUN	… breast cancer …	"breast cancer" is-a "cancer"
NP {,} including {NP ,}* {or | and} NP	… mental disorders including schizophrenia	"schizophrenia" is-a "mental disorder"
such NP as {NP ,}* {or | and} NP	… such digestive disorders as diarrhea, and gastritis	"diarrhea" is-a "digestive disorder" "gastritis" is-a "digestive disorder"
NP {,} such as {NP ,}* {or | and} NP	… cancers such as breast cancer, and lung cancer	"breast cancer" is-a "cancer" "lung cancer" is-a "cancer"
NP {,} especially {NP ,}* {or| and} NP	… blood diseases, especially anemia, and leukaemia	"anemia" is-a "blood disease" "leukaemia" is-a "blood disease"

3.3 Topic Selection

The unsupervised and shallow nature of pattern-based approaches, natural language ambiguity and the unreliability of individual observations may hamper the quality of the retrieved concepts as domain subtopics [26]. In general, when using unsupervised learning techniques, it is necessary to use methods to assess the suitability of the extracted knowledge. In our case, once a set of topic candidates has been retrieved, we need to select the most reliable ones.

The use of *statistical measures* to assess the semantic relatedness between concepts has proven to be an appropriate unsupervised technique [22]. In our case, term co-occurrence in a document set can be used as a measure of the topic candidate relevance in the input domain [11]. Collocation measures are typically used for that purpose [26]:

$$c_k(a,b) = \frac{p(ab)^k}{p(a)p(b)} \tag{1}$$

In this formula $p(a)$ is the probability of the word a occurring within the text and $p(ab)$ the probability of co-occurrence of words a and b. From this formula, one can define the *Symmetric Conditional Probability* (SCP) [10] as c_2 and the *Pointwise Mutual Information* (PMI) [12] in the form $log_2 c_1$.

However, statistical techniques perform poorly when the words are relatively rare, due to the scarcity of data. In our case, the scope of the medical library may not be enough to extract robust statistical values to sustain the semantic assessment of subtopic candidates. Some authors [4] have demonstrated the convenience of using a large amount of texts to improve the quality of classical statistical methods. Concretely, Turney [29] proposed methods to address the sparse data problem by using the hugest available data source: the Web.

The problem is that the analysis of such an enormous repository for computing co-occurrence measures is, in most cases, impracticable. Fortunately, the availability of massive Web Information Retrieval tools can help in that purpose. In fact, it has been claimed that the probabilities of Web search engine terms, conceived as the frequencies of page counts returned by a search engine divided by the number of indexed pages, approximate the relative frequencies of those terms as actually used in society [6].

Following this premise, Turney [29] adapted PMI to approximate term probabilities from Web search hit counts. He defined a score (2) to compute the collocation between an initial word (*problem*) and a related candidate concept (*choice*) in the form:

$$Score(choice, problem) = \frac{hits(problem \ AND \ choice)}{hits(choice)} \qquad (2)$$

Compared to the original PMI formula, since this score is meant to rank a set of *choices* –or candidates- corresponding to a given *problem*, it drops log_2 and *hits(problem)* in the denominator because it is a common value for all choices. The total number of indexed pages (which should divide the hit counts in order to obtain term probabilities instead of absolute occurrences) is also omitted as it is common for the numerator and the denominator.

Following those premises, we use a Web-based statistical approach to assess the reliability of candidate topics against the whole Web, through appropriate queries performed over a Web search engine such as Live Search. We prefer to use the Web instead of local statistics which may be computed from the hit count of the library's search engine due to its much higher size and heterogeneity, which help to obtain more robust and reliable statistics [4]. In this manner, we are able to obtain very robust measures in an immediate manner involving, in most cases, an order of magnitude much higher than the concrete library (e.g. querying *gastritis* in PubMed returns around 20,000 resources; querying it in Live Search, we obtain more than 2,000,000).

Based on Turney's score, we composed a formula (3) which can be used to assess the degree of taxonomic similarity between the domain concept and each subtopic candidate.

$$Score(candidate) = \overset{\#Patterns}{\underset{i=1}{Max}} \left(\frac{hits("Pattern_i("concept","candidate")")}{hits("candidate")} \right) \qquad (3)$$

This score uses the pattern itself as part of the query, joining it to the keyword and the candidate with double quotes (e.g. hits("*blood disorders such as anaemia*")). This kind of queries helps to ensure that the matching count reflects the taxonomic relationship between the two concepts. After extensive experimentation [27], we concluded that highly contextualized queries result in better estimations as they minimize ambiguity problems. This approach is different to other Web-based statistical strategies [29][11] in which only the degree of co-occurrence between words in the text is computed and for which the kind of semantic relationship inherent to the co-occurrence cannot be assessed.

Each of the defined patterns (as presented in Section 3.2) results in a different query and a different score. The final score for each candidate is the maximum of each individual pattern. With this mechanism we try to avoid any constraint or dependence between the candidate and the use of a particular pattern (i.e. the fact of finding a pattern matching in the text for a candidate does not imply that this is the most common expressive form). This formula represents the maximum probability of finding any taxonomic-like construction involving the candidate topic and the input domain in the scope of the Web resources containing that candidate.

Those topic candidates that exceed an empirically set threshold are finally selected. The threshold configures the behaviour of the system in terms of potential precision and recall of the results and it is set to perform accurately for medical domains. A minimum number of hits for the constructed queries is also required in order to avoid misspelled terms.

3.4 Incremental Learning

At this point, the system has to decide if the set of selected topics is enough to have a clear structure of the domain or a deeper analysis should be performed considering additional medical documents in order to offer a representative structure of topics for the domain. The amount of resources needed to have a good domain recall depends on many factors, like the domain's generality or the library's coverage for that area. Due to the automatic and unsupervised nature of our proposal, this parameter cannot be set a priori, so we need a mechanism that regulates the process, providing feedback about how the learning for a particular domain is evolving in order to decide whether to continue evaluating resources or not.

In order to tackle this problem, as shown in Figure 1, we propose an incremental methodology in which the amount of resources analysed at each learning step is increased until most of the topics for the domain have been acquired. More concretely, after the selection procedure is performed, the percentage of selected topics from the list of extracted candidates is computed (4). If it is high, this indicates that the domain is particularly wide and a deeper analysis will potentially return more topics. In this case, we query again the digital library search engine to obtain the next set of resources and repeat the full process.

$$LearningRate = \frac{\#Selected\_topics}{\#Total\_candidate\_topics} \tag{4}$$

The process is iteratively executed until the global percentage of selected topics, computed from the accumulation of results of each iteration, falls below a certain threshold, indicating that most of the categories for the domain have already been acquired and the topics of the remaining documents are potentially redundant. The process also finishes if during 10 iterations we are not able to retrieve any new candidate.

Using this feedback mechanism we ensure the correct finalization of the automatic algorithm regardless of the generality of the explored domain.

4 Using the Topic Hierarchy to Index Medical Documents

At the end of the learning process, we obtain a topic categorization for the domain (e.g. *blood disorder*). Each subtopic represents a concept specialisation. Querying those terms into the repository's search engine, we are able to retrieve resources corresponding to that specialisation (e.g. *anaemia*). Thus, the topic hierarchy (which would likely have to be composed off-line) can be stored on the server side as a structured index of documents which can be easily reused in equivalent user queries. In this manner, the user is able to browse resources in a comprehensive directory-like fashion. The important point here is that topic hierarchies have been obtained automatically, without human supervision and depending on the actual contents of the documents available in the repository (not using predefined ontologies). So, we complement the functionality of the keyword-based search engine and overcome its main limitations (mentioned in the introduction), derived from its lack of semantics.

As shown in Figure, 1, for each subtopic of the hierarchy the whole learning process can be repeated recursively, obtaining a more detailed multi-level hierarchy (*macrocytic anaemia* is a subtopic of *anaemia* which is a type of *blood disorder*). In this way the user can request further details in the particular topics in which he/she is particularly interested.

The proposed methodology has been implemented as a Web interface that is placed on top of the PubMed digital library and provides a way to access its resources in a directory-like fashion (see Figure 2). In this manner, the system controls the access to the library's search engine via the browsing of the presented hierarchy, which is translated into queries to the search engine, retrieving the appropriate resources in a transparent way.

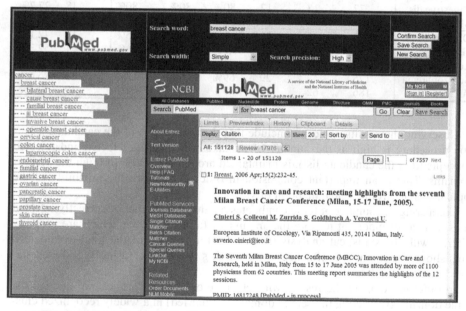

Fig. 2. Web interface provided by the system for the PubMed electronic library

5 Evaluation

In order to evaluate the quality of the obtained topic hierarchies, we have checked the correctness of the is-a relationships between the concepts. However, as the hierarchy is constructed in function of the coverage of the PubMed library for that domain, we cannot compare it to a general-purpose domain taxonomy (e.g. the taxonomy of the *anaemia* concept in UMLS or MeSH). So, we requested two human experts to perform a *concept-per-concept evaluation* [25] of the topic candidates extracted in several iterations of the analysis, checking their validity as domain subtopics. Comparing them against the selected or rejected candidates returned by our statistically-based selection procedure we can compute the performance of the proposed algorithm, using the standard scores of *recall, precision and F-Measure*.

As evaluation test, we present the results obtained for several medical domains covered in the PubMed electronic library. We searched for several high-level medical concepts (such as *blood* or *digestive diseases*) and then we refined some of the obtained subtopics (e.g. *gastritis* and *diarrhoea* are subtopics of *digestive disease, leukaemia* and *anaemia* of *blood disease* and *schizophrenia* of *mental disorder*). The evaluation results are presented in table 2.

Table 2. Evaluation results for medical domains in the PubMed digital library for our approach and the Clusty search engine (the two columns on the right)

Domain	#Topic candid.	#Selec. topics	#OK topics	Precis.	Recall	F-Meas.	#Clusty Topics	Clusty Precis.
Blood disease	105	13	11	84,5%	78,5%	81,4%	4	25%
Digestive dis.	72	15	12	80%	80%	80%	18	27,7%
Mental dis.	102	20	16	80%	80%	80%	17	29,4%
Diarrhoea	192	71	52	73,2%	94,5%	82,5%	19	42,1%
Gastritis	91	15	13	86,6%	86,6%	86,6%	21	52,3%
Leukaemia	156	52	38	73%	84,4%	78,3%	15	46,6%
Schizophrenia	175	42	29	69%	93,5%	79,4%	15	26,6%
Anaemia	186	62	44	70,9%	86,2%	77,8%	19	42,1%

Observing the results, we can conclude that the correctness of the candidate selection procedure is high as the number of mistakes (incorrectly selected and rejected concepts from the candidate list), is maintained around a 15-30% through the various domains. This is an important conclusion in order to measure the reliability of the topic hierarchy from a semantic point of view.

Comparing the number of topic candidates against the number of correctly selected ones, it is also important to note the amount of false candidates which are properly filtered with the statistical analysis (from a 75% to a 90% of the total candidates). This shows the important influence in the result's precision of using Web-based statistics to filter candidates.

In order to compare the results with other automatic approaches, we queried the same domains (restricting the search domain to PubMed) in a widely recognized cluster-based search engine: Clusty (considered as one of the best available systems [14]). Clusty results are evaluated in the last two columns of Table 2.

One can see that Clusty classifications tend to return a lower amount of topics with a very low precision. From a qualitative point of view, the returned categories suffer from a poor structure, resulting in an arbitrary number of topics with a very diverse degree of generality. Unlike our approach, from the semantic point of view, Clusty's categories lack consistency, as different relationship types (specialisations, domains of application, institution names, barely related concepts, etc.) are merged. This hampers the domain comprehension and makes difficult the exploration and navigation of the available resources.

6 Conclusion and Future Work

The methodology presented in this paper allows the automatic creation of topic hierarchies in order to bring taxonomic structure to electronic repositories. The approach exploits well-known knowledge acquisition techniques, like linguistic patterns and statistical analysis. Although those techniques are domain-independent, our unsupervised approach is especially suited to dynamic, knowledge-intensive and technical domains like Medicine, which are very scarcely represented in typically used ontologies like WordNet. As presented in the previous section, the system provides better and more comprehensive structures than cluster-based approaches.

The resulting topic hierarchies can bring interesting benefits to the users of an electronic repository. On the one hand, they allow non-specialised users to browse and access the library's electronic resources in a directory fashion in an intuitive way. On the other hand, the system could also represent a valuable tool for Web masters or domain experts that can automatically generate indexes for large digital libraries. It is important to note that the fact that hierarchies are constructed from scratch allows accurately represent the actual topics covered by document repositories, unlike more general solutions employing predefined domain ontologies to ease Web browsing.

An interesting line of future work is the possibility of using the system to perform automatic semantic annotation of the library's resources (e.g. annotating the documents with references to the concepts of the constructed hierarchy). This idea would be a step forward towards the Semantic Web [2], and could lead the way towards performing automated reasoning procedures on the knowledge of the library.

In addition, as synonymia is very common in Medicine, we plan to incorporate the detection of synonyms or lexicalizations for the obtained taxonomies. Preliminary research about this issue is reported in [28].

From the evaluation point of view, it would also be interesting to test the use of the created taxonomies for navigation with expert and novice domain users. Another line of future work is the investigation of whether the results can be improved if a background taxonomy (such as MESH) is used as the starting point of the learning process, instead of starting from scratch.

Acknowledgements

The work has been partially supported by the Universitat Rovira i Virgili (2009AIRE-04), the K4Care European research project (IST-2004-026968) and the HYGIA Spanish project (TIN2006-15453-C04-01). The authors would also like to acknowledge Rubén Soriano for the implementation of some system modules.

References

1. Agirre, E., Ansa, O., Hovy, E., Martínez, D.: Enriching very large ontologies using the WWW. In: Proceedings of the Workshop on Ontology Construction of the European Conference of AI, Berlin, Germany (2000)

2. Agrawal, R., Imielinksi, T., Swami, A.: Mining association rules between sets of items in large databases. In: Proceedings of the ACM SIGMOD Conference on Management of Data, pp. 207–216 (1993)
3. Berners-Lee, T., Hendler, J., Lassila, O.: The semantic web. Scientific American 5(284), 34–43 (2001)
4. Brill, E., Lin, J., Banko, M., Dumais, S.A.: Data-intensive Question Answering. In: Proceedings of the Tenth Text Retrieval Conference, pp. 393–400 (2001)
5. Chung, C.Y., Lieu, R., Luk, A., Mao, J., Raghavan, P.: Tematic Mapping – From Unstructured Documents to Taxonomies. In: Proceedings of the 11th International Conference on Information and Knowledge Management, USA, pp. 608–610 (2002)
6. Cilibrasi, R., Vitanyi, P.M.B.: The Google Similarity Distance. IEEE Transactions on Knowledge and Data Engineering 19(3), 370–383 (2006)
7. Cimiano, P., Staab, S.: Learning by Googling. Proceedings of SIGKDD Explorations 6(2), 24–33 (2004)
8. Ciravegna, F., Dingli, A., Guthrie, D., Wilks, Y.: Integrating Information to Bootstrap Information Extraction from Web Sites. In: Proceedings of the IJCAI Workshop on Information Integration on the Web, pp. 9–14 (2003)
9. Cutting, D., Karger, D., Pedersen, J., Tukey, J.W.: Scatter/Gather: A Cluster-based Approach to Browsing Large Document Collections. In: Proceedings of the 15th Annual International ACM/SIGIR Conference, Copenhagen, pp. 318–329 (1992)
10. da Silva, J.F., Lopes, G.P.: A local maxima method and a fair dispersion normalization for extracting multi-word units from corpora. In: Proceedings of Sixth Meeting on Mathematics of Language, pp. 369–381 (1999)
11. Etzioni, O., Cafarella, M., Downey, D., Popescu, A.M., Shaked, T., Soderland, S., Weld, D.S., Yates, A.: Unsupervised named-entity extraction from the Web: An experimental study. Artificial Intelligence 165, 91–134 (2005)
12. Fano, R.: Transmission of Information. MIT Press, Cambridge (1961)
13. Fensel, D.: Ontologies: A Silver Bullet for Knowledge Management and Electronic Commerce. Springer, Heidelberg (2001)
14. Freeman, R.T.: Topological Tree Clustering of Web Search Results. In: Corchado, E., Yin, H., Botti, V., Fyfe, C. (eds.) IDEAL 2006. LNCS, vol. 4224, pp. 789–797. Springer, Heidelberg (2006)
15. Gómez-Pérez, A., Fernández-López, M., Corcho, O.: Ontological Egineering, 2nd edn. (2004)
16. Grefenstette, G.: SQLET: Short Query Linguistic Expansion Techniques: Palliating One-Word Queries by Providing Intermediate Structure to Text. In: Proceedings of Information Extraction: A Multidisciplinary Approach to an Emerging Information Technology, RIAO 1997, Montreal, Canada, pp. 97–114 (1997)
17. Hahn, U., Schulz, S.: Towards Very Large Terminological Knowledge Bases: A Case Study from Medicine. In: Proceedings of Canadian Conference on AI, pp. 176–186 (2000)
18. Hearst, M.A.: Automatic acquisition of hyponyms from large text corpora. In: Proceedings of 14th International Conference on Computational Linguistics, France, pp. 539–545 (1992)
19. Ismond, K.P., Shiri, A.: The medical digital library landscape. Online Information Review 31(6), 744–758 (2007)
20. Kietz, J.U., Maedche, A., Volz, R.: A Method for Semi-Automatic Ontology Acquisition from a Corporate Intranet. In: Proceedings of the EKAW 2000 Workshop on Ontologies and Texts, Amsterdam, The Netherlands. CEUR Workshop Proceedings, vol. 51, pp. 4.1–4.14 (2000)

21. Kobayashi, M., Takeda, K.: Information Retrieval on the Web. ACM Computing Surveys 32(2), 144–173 (2000)
22. Lin, D.: Automatic Retrieval and Clustering of Similar Words. In: Proceedings of the 17th International Conference on Computational Linguistics and 36th Annual Meeting of the Association for Computational Linguistics, Montreal, Canada, pp. 768–773 (1998)
23. Maarek, Y.S., Fagin, R., Ben-Shaul, I.Z., Pelleg, D.: Ephemeral document clustering for Web applications. Technical Report RJ 10186, IBM Research (2000)
24. Morin, E.: Automatic acquisition of semantic relations between terms from technical corpora. In: Proceedings of the fifth international congress on terminology and knowledge engineering. TermNet-Verlag, Vienna (1999)
25. Navigli, R., Velardi, P.: Learning Domain Ontologies from Document Warehouses and Dedicated Web Sites. Computational Linguistics 30(2), 151–179 (2004)
26. Popescu, A., Etzioni, O.: Extracting Product Features and Opinions from Reviews. In: Proceedings of the conference on Human Language Technology and Empirical Methods in Natural Language Processing, Vancouver, Canada, pp. 339–346 (2005)
27. Sánchez, D., Moreno, A.: Pattern-based automatic taxonomy learning from the Web. AI Communications 21(1), 27–48 (2008)
28. Sánchez, D., Moreno, A.: Automatic Discovery of Synonyms and Lexicalizations from the Web. In: Artificial Intelligence Research and Development, pp. 205–212. IOS Press, Amsterdam (2005)
29. Turney, P.D.: Mining the Web for synonyms: PMI-IR versus LSA on TOEFL. In: Flach, P.A., De Raedt, L. (eds.) ECML 2001. LNCS (LNAI), vol. 2167, pp. 491–499. Springer, Heidelberg (2001)
30. Zamir, O., Etzioni, O.: Grouper: a dynamic clustering interface to Web search results. In: Proceedings of the Eighth International WWW Conference, Canada, pp. 1361–1374 (2000)
31. Zhang, D., Dong, Y.: Semantic, Hierarchical, Online Clustering of Web Search Results. In: Proceedings of the 6th Asia Pacific Web Conference, China (2004)

Bridging an Asbru Protocol to an Existing Electronic Patient Record

Claudio Eccher[1], Andreas Seyfang[2], Antonella Ferro[3],
Sergey Stankevich[1], and Silvia Miksch[2]

[1] Fondazione Bruno Kessler, Trento, Italy
{cleccher,stankevich}@fbk.eu
[2] Danube University Krems, Austria
seyfang@ifs.tuwien.ac.at, silvia.miksch@donau-uni.ac.at
[3] Medical Oncology, S. Chiara Hospital, Trento, Italy
antonella.ferro@apss.tn.it

Abstract. Clinical protocols can improve the quality of care if implemented in Decision Support Systems (DSS) that are used in clinical practice. For optimal user acceptance, they must use data from the existing Electronic Patient Records (EPR) and enforce only small changes in the care process and minimal extra effort for data entry. In this paper we describe how we handle the challenge of mapping a breast cancer treatment protocol encoded in Asbru to a legacy EPR which has been used by oncologists at the point of care for years. We identified different levels of integration effort ranging from readily available data in the EPR to abstractions which can only be performed by domain experts. By involving the author of the protocol in the implementation process, we were able to design a system which promises to improve the daily routine at the places of application.

1 Introduction

Modeling guidelines and protocols in machine-processable form in a Decision Support System (DSS) allows to support the treatment of patients according to predefined rules. It increases the adherence to these rules and reduces the workload if the system is well accepted in practice. To this end, it is crucial that the DSS is integrated in the dataflow and the existing user interfaces at the point of care [1,2].

In this paper, we describe our work within the Oncocure project at the Medical Oncology Unit (MOU) of the S. Chiara Hospital of Trento (Northern Italy), mapping a breast cancer protocol encoded in Asbru to the legacy Electronic Patient Record (EPR) used at the MOU. In the course of the analysis we identified different settings: (1) data directly available in the EPR; (2) abstractions computable from clinical data; (3) abstractions performed by the clinician, which either include 'holistic' assessments or abstraction rules that cannot be formalized with reasonable effort; (4) proposed extensions to the data stored in the EPR. We also discuss how the analysis of the EPR changed our understanding of details in the protocol, and how missing values are handled.

D. Riaño et al. (Eds.): KR4HC 2009, LNAI 5943, pp. 14–25, 2010.

1.1   The Legacy Oncologic Electronic Patient Record

The oncologic EPR [3], designed and developed in strong collaboration with the end users, is a Web-based application for the management of the care process for oncologic patients, which allows oncologists to collect, manage, and visualize clinical data stored in a relational database. The EPR contains a set of temporally ordered clinical data regarding observations (e.g., laboratory results) and medical acts (e.g., surgeries, past and ongoing therapies), collected by the oncologists during an encounter, which is any daily contact of a patient with the MOU: an outpatient visit for consultation; a one-day stay for therapy administration; or a day of hospitalization for severe problems. The presentation of data is organized in virtual folders according to the care delivery path in an oncologic ward (anamnesis, therapy administration, follow-up). A considerable effort was made to codify as many clinical data as possible, in order to allow their later reuse for statistical analysis or automatic extraction.

Initially deployed in 2000 in the MOU, the EPR was subsequently shared with the Radiotherapy Unit and the Internal Medicine Units of several peripheral hospitals. Up to now, it contains data of more than 12,000 patients. Breast cancer is by far the most common disease, amounting to about 4,000 cases.

1.2   The Oncocure Project

The cancer care process carried out at the MOU is centered around periodic encounters with the patient, in which the oncologist decides the appropriate therapeutic strategy on the basis of the patient visit and exam results. The decision is based on internal protocols: local adaptations of national and international guidelines, integrated with knowledge from other sources (e.g., consensus conferences), prepared by the oncologist specialist for the corresponding cancer type.

The two-year Oncocure project, started in 2007, aims at designing and developing a prescriptive guideline-based DSS integrated with the legacy oncologic EPR in use at the MOU, based on the Asbru model of the internal protocol for the administration of medical therapy to breast cancer patients. The DSS will provide the oncologists with the most appropriate therapeutic strategy for the given disease and patient condition.

2   Related Work

The problem of integrating computerized guidelines with legacy systems into the clinical workflow has been attacked by the Medical Informatics Community for years. Several authors have discussed the idea of a Virtual Medical Record (VMR) middle layer, either based on some standard model [4] or on commercial platforms [5]. In [6], the authors deal with the problem of integrating a DSS for supporting stroke prevention and management with a proprietary text-based EPR. Besides using a middle layer, they agreed with the EPR developers to change the EPR from free text to ICD9-CM encoded data, where needed; this step was not necessary in our case, since the oncologic EPR already uses domain terminologies for the relevant data.

In [7], the benefits of the full integration of guideline management components in the information system architecture of a healthcare organization are analyzed and a methodology based on the European prEN12967 standard to facilitate this integration is proposed.

The SAGE project [8] aimed at creating a demonstration methodology and infrastructure to integrate guideline-based decision-support technology in commercial clinical information systems, through a deployment-driven methodology involving identification of usage scenarios, distillation and disambiguation of relevant guideline knowledge, formalization of guideline data elements and vocabulary, and encoding of scenarios and guideline knowledge in an executable model.

Recently, Mor Peleg [9] described the application of a GLIF based DSS for diabetes to an existing EPR. In her case, it was necessary either to remove unavailable data items from the guideline or to modify the EPR schema. In our case we were more lucky, since the protocol we modeled was designed at the place of application, and because we could ensure the additional entry of the missing inputs.

3   Modeling Aspects

Most of the knowledge needed for a smooth integration of the DSS and the EPR is not in the protocol, but is part of the tacit domain and care process knowledge possessed by oncologists, which we were able to explicit only after long discussions with them. After a phase of analysis of breast cancer protocol and EPR content, we approached the mapping of Asbru and EPR by grouping the parameters used by the Asbru model according to different scenarios of integration.

3.1   Levels of EPR Data Integration

We were able to define four different levels of integration. First of all, in few cases we could use the data in the EPR as they are. Secondly, we identified a group of parameters resulting from the automatic abstraction of the required information from data in the EPR. In several cases, the process of defining these abstractions even contributed to a better understanding of the protocol. Thirdly, we identified a group of parameters for which the abstractions contained subjective assessments by the physician in addition to objective abstraction rules, and we opted for leaving the parameter assessment to the physician, to reduce data entry. Moreover, data in the EPR may not be reliably complete; thus, the matching could not be performed automatically. Forth and finally, a set of values is considered candidate to be stored in the EPR in the future, after a time of evaluation.

Data directly available in the EPR. Notwithstanding the protocol was developed at the site of daily use of the EPR, the number of data items in the protocol which are directly available in the EPR was found to be surprisingly small.

On one hand, the EPR was conceived to store *low-level* data. They can be results of diagnostic tests (e.g., hormone receptor status measured by pathologists on the removed tissue), or therapy administration data (e.g., drugs, period of administration, etc.). On the other hand, the protocol eligibility conditions are based on abstractions made by the oncologist at the point of care, using the clinical data stored in the EPR (e.g., hormone responsiveness, based on hormone receptor status).

We were able to identify only four parameters directly corresponding to EPR data: namely, the histological grade G, which defines the grade of differentiation of tumor cells; the lymphatic invasion L, which expresses the penetration of lymphatic transport conduit by a malignancy; the percentage of Mib-1 positive receptors; and the regional lymph node involvement N, a categorical value part of the TNM classification system [10], that represents the extent of tumor spreading to regional lymph nodes. Regional lymph node involvement can be assessed by the oncologist during a patient visit (clinical N) or by the pathologist on the surgically removed lymph nodes (pathological N). In the EPR, the first value recorded is the clinical N, which is substituted with the pathological N when the pathology laboratory results are available.

We identified a second set of clinical data in the EPR not directly used in the protocol but needed for the abstractions described below: the tumor extension T, a categorical value of the TNM classification system; the percentage of progesterone and estrogen receptors (PgR+ and ER+); a four-valued score defining the degree of expression of the Her2 gene (c-erb2); and the result of the *FISH* (Fluorescence In Situ Hybridization) test.

Parameters automatically abstracted from the EPR. This set of parameters is constituted by data that are not directly stored in the EPR, but that can be computed without human intervention from lower-level data using clearly defined rules. The abstractions could be performed by the Asbru data abstraction module, by defining the rules in the model. However, after extended discussions, we opted to move most of them in an external rule engine mainly for two reasons:

- While the abstractions are, in general, rather stable, the rules and parameters used for their computation may require more frequent revisions. This choice makes the maintenance of the guideline model easier;
- We can keep the number of parameters which are transferred from the EPR to the Asbru interpreter low, thus optimizing the performances of the system.

The computation algorithms can be simple arithmetic operations. For example: the patient age, simply calculated as the difference between the date of first diagnosis of cancer or of the first surgery for removing the tumor mass, if performed, and the birth date of the patient; the Disease Free Interval (DFI), computed as the difference between the date of the surgical removal of the tumor and that of the first event of local or distant recurrence (metastasis). In other cases, the rules can be set of complex logical conditions, whose truth value determines the value of the abstraction. For example:

- Hormone responsiveness (HR) is a three-valued abstraction (certain, uncertain, absent), which is computed by the logical combination of conditions involving ER+, PgR+, Mib1, G, N, L, and T;
- HER2 positiveness expresses whether the Her2 gene in tumor cells is over-expressed or not. It can be computed from c-erb2 and FISH.

Therapy history parameters. Decision of treatment of local recurrences and distant metastasis depends on the use of specific classes of drugs in previous treatment phases. They are specified in the protocol as Boolean parameters, expressing presence/absence of a drug of a certain class. Examples are taxanes or anthracyclines in the adjuvant

chemotherapy, aromatase inhibitors in adjuvant hormone treatment, Tamoxifen in previous metastatic treatment lines. The evaluation of these parameters requires the definition of temporal abstractions, that we perform outside the model.

The entire history of past therapies, and hence the temporal abstractions above, in
fact, can be automatically reconstructed from the EPR, which records the details of
administered therapies: the kind of therapy (hormone, immuno- or chemotherapy), its
intent (adjuvant, neoadjuvant, palliative), the sequence of regimens that compose the
therapy (e.g., FEC followed by TXT), the cycles of each regimen (e.g., FEC × 3 cycles),
the drugs of each regimen (e.g., Fluorouracil, Epirubicin, Cyclophosphamyde), the parts
of the regimen (e.g., Part 1, Fluorouracil on day one), the standard dose and the possible
dose reduction of each drug, either decided in advance or during the treatment cycle.
Since in the EPR references to the active principle in a drug are mixed with references
to commercial product names, we defined an external ontology of the oncologic drugs
mentioned in the protocol, based on the NCI thesaurus [11].

Abstractions requiring user assessment. In several places, the protocol refers to patient and disease states in a rather holistic fashion. Such expressions could be broken
down into elementary findings and combination rules for them. However, the elements
of the resulting expressions would themselves require subjective judgment from the side
of the physician. Furthermore, for some of these entities, the abstraction rules cannot be
acquired and represented with certainty. In these cases, it is more efficient to ask only
one question to the doctor. The set of these "holistic" parameters includes:

Eligibility for a certain treatment. These are boolean parameters which define the eligibility of a patient for a (class of) treatment. For example, the possibility of administering
chemotherapy depends not only on medical circumstances, i.e., whether the patient is
physically fit for chemotherapy, which in any case requires the assessment of the doctor. It is also prohibited during pregnancy. Furthermore, the consent of the patient to
undergo this treatment is required.

The eligibility for clinical-trials, which are multi-centric studies to test the effectiveness of new medical therapies, implies the fulfillment of enrollment criteria more rigid
and better detailed than those of normal treatments. First of all, the decision to enroll a
patient may not be taken by an oncologist alone, but may require a collective discussion.
Secondly, the explicit consent of the patient to participate in the trial is required.

Aggressiveness of the disease. The rule for determining the aggressiveness of the tumor,
required to decide the metastatic treatment, is not (and cannot) be precisely defined in
the protocol, being a complex combination involving both objective data (e.g., positive
hormone receptors) and definitely holistic data (e.g, low disease load, good response to
previous treatments).

Contra indications to some (classes of) drugs. Specific oncologic drugs or classes of
drugs (e.g, anthracyclines) have relative or absolute contraindications. The former allow
the administration of a drug, upon careful medical judgment and/or continuous control,
the latter does not allow the drug administration at all. The protocol includes general
contraindication statements (e.g., contraindications to anthracyclines) without further
specification. Actually, the protocol does not specify the contraindications, which are

part of the tacit knowledge of the physicians. We were able to explicate them after discussion with the oncologist, which allowed us to divide them in two groups:

- Contraindications related to the conditions of the patient at the time of decision making, for example a pregnancy or a desire of pregnancy that forbids the administration of Tamoxifen. These parameters, of course, are not permanently registered in the EPR and must be acquired by the oncologist.
- Contraindications due to the presence of specific past or concomitant diseases: e.g., a past history of endometrial polyps for Tamoxifen, a cardiomyopathy for anthracyclines and Herceptin. These data are not part of the minimum data set necessary for the main treatment decision, whose collection is enforced by rules implemented in the EPR, and the oncologist may have not collected them for various reasons. Hence, the absence of related diseases cannot be deducted from the absence of these data. The strategy we adopted is to set the parameter to yes, if a disease is found in the EPR, otherwise, we require the user to enter a concise value (yes/no).

Postmenopause. Doctors carefully consider the onset of menopausal state. When a chemotherapy is given, the treatment can cause amenorrhea, more or less prolonged or even irreversible (iatrogenic menopause). After chemotherapy, the decision of giving some hormone treatments (e.g., aromatase inhibitors, ovarian function suppressors) depends on the presence of a true menopausal state.

Variables to be added to the EPR. After modeling the protocols, we have two kinds of data required by the protocol that are not or may not be available in the EPR.

- Data that the EPR allows to collect, but the collection of which is not guaranteed. Comorbidities for contraindicating some treatment belong to this group. In this case we can i) ask for implementation of additional rules to the EPR enforcing their collection, or ii) explicitly ask to oncologists to collect a minimum set of comorbidities as part of the routine anamnesis assessment.
- Data that cannot be currently collected in the EPR, which correspond to the abstractions requiring user assessment. This is probably due to the fact that oncologists did not feel the need to directly collect them because they use their knowledge to abstract them from lower level data.

It would be advantageous to add to the EPR at least the set of parameters that, once assessed, are valid forever: post menopause, eligibility for a certain clinical trial, aggressiveness of the metastatic disease and absolute contraindications to some therapy. Since this step has complex organizational aspects, however, we decided to maintain them in a separate repository. After a phase of initial system use and evaluation, the oncologists will consider the opportunity to ask EPR developers to extend the EPR to accommodate some of these parameters.

3.2   Related Issues

Better understanding of the protocol from the EPR. There is a set of values to which the protocol refers like independent variables, some of them numerical, but in fact they are qualitative values of one datum in the EPR.

Tumor extension. Several treatments depend on a threshold on the size of the tumor, measured in cm. Usually, however, oncologists use the categorical T scale, part of the TNM staging system to express the tumor dimension. For breast cancer, the categories from T0 to T3 (and their subcategories) classify the tumor according to its maximum dimension. The category ranges coincide with the thresholds in the protocols. Moreover, the involvement of adjacent structures (skin and thoracic walls), along with the inflammatory nature of the tumor, are conditions suggesting neoadjuvant therapy. These conditions are indicated by assigning the tumor to the T4 and T4d categories, respectively. Thus, we concluded that tumor size is used as a qualitative value following the TNM scheme rather than a quantitative value, even though the notation in the protocol (e.g., < 2 cm) seems to suggest the latter.

Disease state. One variable in the EPR represents the general assessment of the disease state by the physician. It is collected during each encounter of the oncologist with the patient. It is a categorical datum, whose possible values are *Not definable, No evidence of disease* (set after a radical surgery usually followed by adjuvant therapy), *Complete Response, Partial Response > 50%, Partial Response < 50%, Mixed Response, Stable disease, Progression.* It is estimated by comparing the tumor size and/or tumor spreading before and after the treatment, resulting from the physical examination and CT scan images. The disease state is fundamental to measure the success or failure of a therapy and determines the choice of the next strategy. In the protocol, this parameter is referred to under many different names, which seem to refer to distinct variables. Examples are *partial response > 50%, responsive to chemotherapy, responsive to hormone therapy.*

Dealing with missing data and X categories. For those data directly available in the EPR, we had to deal with two potential problems, which could impair the correct functioning of the DSS:

- Some values could be missing because the oncologist did not record them.
- Several categorical tumor staging parameters used by the oncologists and pathologists, namely TNM staging, vascular invasion L and grading G, include a category X, properly used to "denote the absence or uncertainty of assigning a given category when all reasonable clinical or pathological maneuvers have been used in staging." [12] Although their use is clinically consistent, the protocol does not cover these categories in any eligibility conditions.

After discussions with the oncologist, we divided the parameters in two sets: those parameters that the oncologist considers mandatory in her decision-making process and those parameters that the oncologist considers less important, because they are used to "refine" the decision.

The former includes the age of the patient, the lymph node involvement N, the tumor extension T, the progesterone and estrogen receptor percentages, the Her2 score and the FISH. Since these parameters determine the first-level categorization of patients for the choice of a therapeutic strategy (hormone therapy, chemotherapy, immunotherapy or a combination of them), the oncologists use particular care in collecting them, and we can assume that the users always know these data, otherwise neither them can take a decision. Moreover, in the usual care process workflow the oncologist stores these data in the EPR before deciding a therapy. If these data are not yet stored in the EPR, however, our system can enforce their collection before giving a recommendation; as a matter of fact this constraint is accepted by the expert. Category X cannot cause a problem either, because for operable invasive breast cancer the tumor extension and lymph node invasion is always determined, while for the metastatic breast cancer decision process T and N are not required.

The second set of parameters includes the vascular invasion L, the Mib1 receptor percentage, and the histological grade G. After deciding the general therapeutic strategy, these parameters contribute to the decision of raising the risk class of the patient, and thus to perform a more aggressive treatment. For this set of parameters we decided to adopt a precautionary principle, usually implicitly adopted by the oncologists of the Medical Oncology Unit: in the presence of a missing or unknown (X) value, we recommend the treatment with less severe impact (in terms of adverse effects for the patient). This corresponds to assign the histological grade and the vascular invasion to the lowest categories (G1 and L0, respectively), and the percentage of Mib-1 receptors to 0. Of course, this is not a general rule, since other healthcare centers could adopt the opposite position, but this is part of the well known process of local adaptation of general clinical practice guidelines.

4 Implementation Aspects

System architecture. The distinction between different types of parameters made above and the strategies for dealing with the various types is the basis for the implementation of the DSS architectural solution, shown in Figure 1. The DSS is composed by several software modules:

Database Manager. The Database Manager interfaces the DSS with the EPR database, implementing the queries to retrieve the data directly available in the database and stores them in the Virtual Medical Record (VMR).

Data Abstraction Manager. The Rule Engine Manager wraps a rule engine that computes the abstractions from the lower level data, because we choose to separate them from the protocol model in Asbru, easing the maintenance of the model itself. We use the freely available rule engine CLIPS [13].

Asbru Manager. The Asbru Manager wraps the interpreter and implements the interfaces to provide the interpreter with the parameters from the VMR, running it and forwarding recommendations and requests to the user.

DSS GUI Manager. The DSS GUI Manager manages the interaction of the DSS with the graphical user interface of the oncologic EPR, sending requests and recommendations and receiving user confirmations and answers.

DSS Coordinator. The DSS Coordinator coordinates the interactions of the other modules, by invoking the implemented methods and passing messages between them.

The architecture of the system allows to plug and unplug the software modules, which communicate through XML-based messages, so that each module can be easily

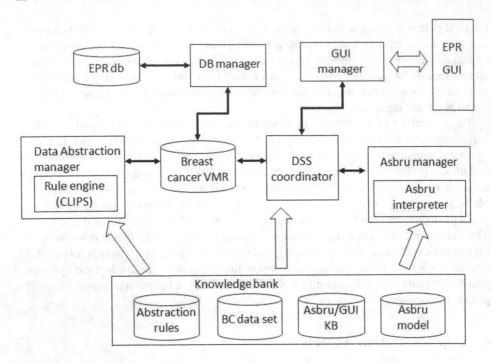

Fig. 1. System architecture of the breast cancer Decision Support System

substituted. For example we could connect to a different database, or substitute CLIPS with a different engine.

A simplified XML-based VMR with a minimum set of objects relevant to our DSS (observations, substance administrations and procedures) contains the data representing the patient clinical history, which are necessary for the Asbru interpreter to take a decision on the treatment. The VMR is divided in three sections: low level clinical data, computable abstractions, and holistic parameters.

A knowledge bank contains the three interrelated knowledge bases for the specific disease (breast cancer, in the current implementation): the Asbru protocol model, used by the Asbru interpreter; the rules for computing the abstractions, used by the rule engine; and the VMR template. The latter is the specification of the minimum dataset that constitutes the VMR for the specific disease (name and type of each parameter, allowed values, range, etc.).

Implementation state. At the time of submitting this paper, the system is being implemented in the Medical Oncology Unit of the S. Chiara Hospital. At the same time, tests for evaluating the correctness of the model and fine tuning the system and the user interface are performed in our laboratory in close collaboration with the breast cancer oncologist. Evaluation of the system acceptance by the medical experts, who represents the users, is scheduled for November and December 2009. After this, a phase of evaluation of system efficacy is planned for the next year.

5 Conclusion

An effective DSS must be integrated with the patient management system at the point of care. Hence, the mapping of the DSS to the EPR in use at the local site may require the adaptation of the clinical protocols and the modification of the EPR. This is not only a matter of building an intermediate layer separating the EPR and the DSS, but first of all a problem of analysis and disambiguation of the knowledge in the protocols, the parameters in the encoded model, and the data available in the EPR.

Protocols are written by oncologists as reference documents that adapt best practice guidelines – integrated with knowledge from other sources – to the local organization, supposed to be used during the daily clinical activity . Consequently, protocols are more concise than guidelines and tailored to the process and resources in the organization.

The legacy EPR was designed and developed in parallel to protocols at the same site as a repository of a minimum data set allowing the oncologists to take informed decision about the best treatment for a patient.

Differently from Mor Peleg's approach [9], which involved the initial encoding only by computer scientists, we worked in close collaboration with the oncologist since the beginning of the encoding activity to deeply understand the care process, how protocol support it, and how the EPR is used in the MOU. This approach, together with the closely related development history of protocol and EPR, facilitated the task of bridging the gap between the DSS and the EPR. In fact, the only modifications to the original protocol model concerned data types, and in some cases the replacement of several Booleans by a single qualitative value. Moreover, all objective data are already present in the EPR, avoiding the burden of heavily modifying the EPR database and interface.

There is some gap between the EPR data and the parameters required by the protocols, mainly because of the amount of implicit knowledge the oncologist applies in abstracting the clinical condition of a patient from EPR data. After explicating this knowledge, we were able to fill part of this gap by automatic abstractions.

The remaining part of subjective assessments by the physician can be seen as violating the principle of standardization of treatment, which is the fundamental idea behind protocols. Unfortunately it cannot be avoided in many cases, since it depends on complex implicit knowledge that can be hardly codified. In the context of our work, it is important that we create a list of such subjective assessments as part of the modeling process. If any of them can be redefined as an abstraction based on clear rules, it is easy to move it to the level of automatic abstractions.

What was unexpected, however, was the high number of parameters which required some subjective assessment of the doctor as part of their abstraction from data otherwise contained in the EPR. In some cases, these were complex clinical decisions that require clinical knowledge outside the protocol. In other cases, the abstraction rule for the value was clear, but it included several subjective assessments as inputs, which are not stored in the EPR. In such cases, we opted for querying the outcome of the abstraction instead of demanding the input of all the contributing factors. This decision was driven by the aim to reduce user action when working with the system, and not by technical feasibility. In all these cases, the oncologist confirmed that the effort of entering the data is justified by the benefit of storing them in the system, easing the management of important information related to the patient cancer history. Hence, this additional data entry is not perceived as a hurdle in the acceptance of the decision support system.

All in all, the close collaboration between the oncologist and the computer scientists since the beginning of the encoding activity has brought a mutual exchange of knowledge that has allowed computer scientists to better understand the clinical domain and adapt the DSS to the specific characteristics of the oncologic activity, and the oncologist to understand the need for changes and the extra effort required by the implementation of an effective DSS; and to have the right expectations regarding the potential, the impact, and the limits of such a tool.

Acknowledgments. This work has been carried out in the context of the project Oncocure, partially funded by the Fondazione Caritro of Trento.

References

1. Sonnenberg, F., Hagerty, C.: Computer interpretable guidelines: where are we and where are we going? Methods Inf. Med. IMIA Yearbook of Medical Informatics 45(Suppl. 1), S145–S158 (2006)
2. Bates, D., Kuperman, G., Wang, S., Gandhi, T., Kittler, A., Volk, L., Spurr, C., Khorasani, R., Tanasi-jevic, M., Middleton, B.: Ten commandments for effective clinical decision support: making the practice of evidence-based medicine a reality. J. Am. Med. Inform. Assoc. 10, 523–530 (2003)
3. Galligioni, E., Berloffa, F., Caffo, O., Tonazzolli, G., Ambrosini, G., Valduga, F., Eccher, C., Ferro, A., Forti, S.: Development and daily use of an electronic oncological patient record for the total management of cancer patients: 7 years' experience. Ann. Oncol. (2008); Epub ahead of print
4. Johnson, P., Tu, S., Musen, M., Purves, I.: A virtual medical record for guideline-based decision support. In: Proceedings of AMIA Symposium 2001, pp. 294–298 (2001)
5. Ciccarese, P., Caffi, E., Boiocchi, L., Halevy, A., Quaglini, S., Kumar, A., Stefanelli, M.: The New Guide Project: Guidelines, Information Sharing and Learning from Exceptions. In: Dojat, M., Keravnou, E.T., Barahona, P. (eds.) AIME 2003. LNCS (LNAI), vol. 2780, pp. 163–167. Springer, Heidelberg (2003)
6. Quaglini, S., Panzarasa, S., Cavallini, A., Micieli, G., Pernice, C., Stefanelli, M.: Smooth Integration of Decision Support into an Existing Electronic Patient Record. In: Miksch, S., Hunter, J., Keravnou, E.T. (eds.) AIME 2005. LNCS (LNAI), vol. 3581, pp. 89–93. Springer, Heidelberg (2005)
7. Román, I., Roa, L., Madinabeitia, G., Millán, A.: Introducing guideline management in the healthcare information system architecture. In: Bos, L., Blobel, B. (eds.) Medical and Care Compunetics 4. Stud Health Technol Inform, vol. 127, pp. 117–124. IOS Press, Amsterdam (2007)
8. Tu, S., Campbell, J., Glasgow, J., Nyman, M., McClure, R., McClay, J., Parker, C., Hrabak, K., Berg, D., Weida, T., Mansfield, J., Musen, M., Abarbanel, R.: The SAGE Guideline Model: achievements and overview. J. Am. Med. Inform. Assoc. 14(5), 589–598 (2007)
9. Peleg, M., Wang, D., Fodor, A., Keren, S., Karnieli, E.: Lessons Learned form Adapting a Generic Narrative Diabetic-Foot Guideline to an Institutional Decision-Support System. In: ten Teije, A., Miksch, S., Lucas, P. (eds.) Computer-based Medical Guidelines and protocols: a Primer and current Trends. Stud. Health Technol. Inform., vol. 139, pp. 243–252. IOS Press, Amsterdam (2008)

10. Greene, F., Page, D., Fleming, I., Fritz, A., Balch, C., Haller, D., Morrow, M. (eds.): AJCC cancer staging manual, 6th edn. Springer, New York (2002)
11. National Cancer Institute, Office for Communication, Center for Bioinformatics: NCI Terminology browser, ftp://ftp1.nci.nih.gov/pub/cacore/EVS/ (last visited 2009, May 18)
12. Greene, F., Brierley, J., O'Sullivan, B., Sobin, L., Wittekind, C.: On the Use and Abuse of X in the TNM Classification. Cancer 103(1), 647–649 (2005)
13. CLIPS: A Tool for Building Expert Systems, http://clipsrules.sourceforge.net/index.html (last visited 2009, September 28)

From Natural Language Descriptions in Clinical Guidelines to Relationships in an Ontology

María Taboada[1], María Meizoso[1], David Riaño[2], Albert Alonso[3],
and Diego Martínez[4]

[1] Department of Electronics and Computer Science, University of Santiago de
Compostela, Spain
maria.taboada@usc.es
[2] Research Group on AI, Rovira i Virgili University, Tarragona, Spain
[3] Hospital Clinic, Barcelona, Spain
[4] Department of Applied Physics, University of Santiago de Compostela

Abstract. Knowledge Engineering allows to automate entity recognition and relation extraction from clinical texts, which in turn can be used to facilitate clinical practice guideline (CPG) modeling. This paper presents a method to recognize diagnosis and therapy entities, and to identify relationships between these entities from CPG free-text documents. Our approach applies a sequential combination of several basic methods classically used in knowledge engineering (natural language processing techniques, manually authored grammars, lexicons and ontologies), to gradually map sentences describing diagnostic and therapeutic procedures to an ontology. First, using a standardized vocabulary, our method automatically identifies guideline concepts. Next, for each sentence, it determines the patient conditions under which the descriptive knowledge of the sentence is valid. Then, it detects the central information units in the sentence, in order to match the sentence with a small set of predefined relationships. The approach enables automated extraction of relationships about findings that have manifestation in a disease, and procedures that diagnose or treat a disease.

Keywords: knowledge engineering, ontologies, UMLS, clinical practice guidelines.

1 Introduction

Research oriented to promote the use of clinical practice guidelines (CPGs) at the point of care, with the aim of improving the quality of health care and reducing costs, is an extensively recognized topic nowadays [1,2]. The proliferation of CPGs and the possibility of using them in many health care activities have increased their value as knowledge resources. Modeling CPGs in a computer-interoperable form is a necessary task to implement them as electronic applications. Formal languages have been developed to represent clinical guidelines in a computer interpretable manner [3,4], such as Asbru [5], EON [6], GLIF [7], Guide [8], Prodigy [9] and PROForma [10]. However, regardless of the benefits

D. Riaño et al. (Eds.): KR4HC 2009, LNAI 5943, pp. 26–37, 2010.
© Springer-Verlag Berlin Heidelberg 2010

they provide to the health care domain, introduction of computerized clinical practice guidelines still presents a serious difficulty. Among others, there are two problems making their adaptation difficult. One is the lack of integration with electronic health records. Agreement on a standard controlled vocabulary may solve, in part, this problem [11]. The second one is the complexity of modeling CPGs. The first phase in the development of a knowledge intensive system is knowledge acquisition in which knowledge engineers and domain experts create a shared model of a domain expert's knowledge. Several tools can be used to simplify this phase, such as AsbruView [12], AREZZO [13], Tallis [14] or Protege [15], among others. Regardless of the considerable volume of research, the modeling process is still a very complex and labor-intensive activity; therefore, methods are needed to automate parts of this activity.

To date, there has been some very relevant work on information extraction from CPG texts. Kaiser et al. [16] have modeled treatment processes by automatically identifying relevant guideline text parts using information extraction methods. Serban et al. [17] simplify knowledge acquisition by automatically extracting control knowledge (such as clinical action decomposition or sequencing) using linguistic patterns, and an ontology that was specifically constructed for this system. On the other hand, many research groups have already exploited the automated entity recognition and relation extraction from patient reports. Many of them have applied knowledge-engineering approaches, using a syntactic parser with domain-specific grammar rules, lexicons and ontologies. The RECIT system [18] combines syntactic and semantic information to extract conceptual graphs expressing the meaning of components in free text sentences. RECIT uses a dictionary encoding medical entities, modifiers and relationships. MedLEE [19] is a NLP-based system successfully used to process reports from different domains (e.g., radiology, pathology, electrocardiography). SeReMeD [20] is a method used to generate knowledge representations from chest X-ray reports, supported by the UMLS. SeReMeD automatically extracts relationships between a leading concept in the sentence and the identified set of modifiers. The relationship types can be non-specific (they are given by the modifier semantics) and specific (they are predefined relationships triggered by the prepositions found in the sentence).

In this paper, we propose an automated method to recognize diagnostic and therapeutic entities, and to identify relationships between these entities from CPG free-text documents. This work is part of the research project HYGIA [1], which promotes the use of intelligent systems in the processes of acquiring, formalizing, adapting, using and assessing knowledge models that describe Care Pathways. Our approach sequentially applies a set of language engineering procedures, terminology resources and syntactic-semantic rules to gradually map relevant clinical contents to an ontology, from natural language input (Fig. 1). Our approach provides a means of automatically expressing the meaning of relevant text sentences in a form that is computationally tractable, providing a valuable mechanism to reduce effort in CPG modeling.

---

[1] http://banzai-deim.urv.net/~riano/TIN2006-15453/

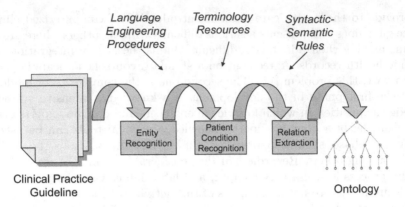

Fig. 1. General steps of our approach

2   Materials

Several resources such as Natural Language Processing (NLP) and Information Extraction (IE) tools, terminologies and ontologies are currently used to support entity recognition and relation extraction [21].

2.1   The Unified Medical Language System

The UMLS [22] provides the terminology resource for the method introduced in this paper. It consists of several knowledge sources providing terminological information. The largest knowledge source is the Metathesaurus, which contains information about medical concepts, synonyms, string-names and the relationships between them. All this information is drawn from over 130 controlled vocabularies, such as SNOMED or MeSH. *Semantic Types* (STs) are a set of basic semantic categories used to classify the concepts in the Metathesaurus. Examples of semantic types are *Diagnostic Procedure* or *Therapeutic Preventive Procedure*. STs are related to each other by hierarchical ('is-a') and non-hierarchical ('diagnoses', 'treats', etc.) relationships, making up the so-called Semantic Network (SN).

2.2   Ontology

As a formal declarative knowledge representation model, an ontology provides a foundation for machine understandable knowledge [23,24]. By providing a comprehensible and formal semantics, ontology helps parts of guideline knowledge in a electronic way to be described, and then these parts can be discovered and analyzed by knowledge engineers and medical experts. Moreover, ontology helps the relationships and properties of concepts to be rigorously defined and linked to guideline chunks. An ontology is a data model that represents a set of concepts within a domain and the relationships among those concepts. By *entity*, we mean some clinical concept or event: diseases, treatments, symptoms and signs, etc. The entity types in our ontology are a subset of the UMLS STs that are necessary to represent concepts relating to diagnosis and therapy recommendations. Diagnosis and treatment entities were considered those with the following STs (and their subtypes): Pathologic Function, Finding, Diagnostic Procedure, Laboratory Procedure, Therapeutic or Preventive Procedure and Pharmacologic Substance. In addition, entities are connected to each other by relationships, such as the 'diagnoses' relationship linking a set of laboratory procedures that are recommended to diagnose a disease, the 'treats' relationship linking a pharmacologic substance that is indicated to treat a disease or the 'has adverse effect' relationship linking a pharmacologic substance with the findings resulting from adverse effects. Some specific examples of sentences describing these types of relationships are shown below. They come from the guideline for the diagnosis and treatment of chronic heart failure (CHF) published by the European Society of Cardiology [25]; they will be used to describe our method in the following section.

Example 1: *Routine diagnostic evaluation of patients with CHF includes complete blood count, S-electrolytes, S-creatinine, S-glucose, S-hepatic enzymes and thyroid function.*

Example 2: *All patients with symptomatic chronic heart failure that is caused by systolic left ventricular dysfunction should receive an ACE-inhibitor.*

Example 3: *Oxygen has no application in CHF.*

2.3   Information Extraction

OpenNLP [2] provides several tools to parse a text into sentences, phrases, lexical elements and tokens. It includes a sentence detector, a tokenizer, a part-of-speech tagger, a Treebank chunker and a Treebank parser.

3   Methods

Our approach applies a sequential combination of several basic methods classically used in knowledge engineering, to gradually map the clinically relevant contents of a natural language input to an ontology. The process consists of several major steps. First, using the standardized vocabulary of the UMLS Metathesaurus, our method recognizes guideline entities. Second, for each sentence, it determines the patient conditions under which the descriptive knowledge of the sentence is valid. Third, it detects the central information units (leading concepts) in the sentence, in order to match them with a small set of predefined relationships.

3.1   Entity Recognition

Our method is built using the OpenNLP toolkit, in order to provide a full parse tree for each sentence selected from the CPG document, and the lexical technique

---

[2] http://opennlp.sourceforge.net/

NormalizeString provided by the UMLSKS API, which is applied to map terms in the text to a set of UMLS concepts. *NormalizeString* normalizes the input terms before searching them in the UMLS database. The normalization process removes lexical differences between strings, such as alphabetic case, inflection, spelling variants or punctuation. The entity recognition process consists of several major sub-steps that are described below.

Selecting relevant information. The clinically relevant contents of the CPG were underlined by a group of physicians taking part in the project. Before extracting sentences from the text, they are annotated with the diagnosis context which they refer to, in order to maintain their context. For example, sentences from the CHF guideline are annotated with *CHF*, or *CHF with PLVEF*[3], or *CHF in elderly patients*.

NLP Parsing. The underlined text is first parsed using OpenNLP modules. Text is tokenized, sentences are found with a sentence splitter, tokens are tagged with part-of-speech labels, and a full parse tree for each sentence is provided by the Treebank parser. Most of the syntactic patterns of the medical entities in the guideline fit a noun phrase (NP). We implemented a NP tagger, which traverses the parse tree and marks up NPs. Our method annotates each NP as either simple, prepositional (Pre-NP) or coordinated (Co-NP). Simple noun phrases do not contain other NP descendants, but Pre-NP and Co-NP usually contain several NPs referring to different medical entities. In later steps, our method processes Pre-NPs and Co-NPs differently from simple NPs, leading to more complex representations in the ontology.

Table 1. Partial results of preprocessing the sentence in example 1

Pre−NP [Routine diagnostic evaluation of patients with CHF]
includes
Co−NP[complete blood count, S-electrolytes, S-creatinine, S-glucose, S-hepatic enzymes and thyroid function]

Mapping noun phrases to medical entities. Each NP is mapped to a set of one or more UMLS concepts using the UMLS NormalizeString service. When there is no mapping for a NP, this is divided into its internal NPs or constituents (i.e., names, adjectives, etc.) and a request is sent to the UMLS database in order to map them to some UMLS concept. Table 2 shows the recovered mappings for example 2.

Disambiguation of medical entities. The ambiguity of names in medicine is very common in natural language, increasing with the use of synonyms and abbreviations. Physicians use polysemy without difficulties, but it is a problem for

---

[3] PLVEF: Preserved Left Ventricular Ejection Fraction.

Table 2. Results of mapping NPs of example 2 to UMLS concepts

String	UMLS concept	Semantic Type
patients	C0030705	Patient or Disabled Group
symptomatic	C0231220	Functional Concept
chronic heart failure	C0264716	Disease or Syndrome
systolic left ventricular dysfunction	C1277187 C1963159	Pathologic Function Finding
ACE-inhibitor	C0003015	Pharmacologic Substance

automated entity recognition. Various disambiguation strategies have been applied to biomedical language processing [26]: established knowledge, supervised and unsupervised learning methods. All this strategies use a corpus for training. Unlike them, our approach apply a set of prioritized heuristic rules, based on syntactic, semantic or terminological criteria:

- Co-NP based rule: If the NP is part of a Co-NP and there is one or more disambiguated NPs, the semantic type of these disambiguated NPs is preferred.
- Semantic types based rule: Some semantic types are preferred to others. In this way, the UMLS semantic types were chosen for creating priority orders. For example, a procedure is preferable to a substance.
- String matching based rule: in case of persistent disambiguation, string similarity between entity names and NPs is preferred.

3.2   Patient Condition Recognition

In this stage, the patient conditions under which the descriptive knowledge of the sentence is valid, are automatically identified. First, the sentence is normalized and then, the patient condition is semi-formally represented.

Normalization. The patient condition is automatically identified by means of a set of regular expressions that are triggered by types of noun or prepositional phrases, such as these:
in patients with *
in patients who *
in <Pathologic Function> [with | from | and | or <Pathologic Function>]
*in * [patients tolerant | intolerant] to <Pharmacologic Substance> **, etc.
where * may be a Finding, a Pathologic Function or a Pharmacologic Substance.

These linguistic regularities occurring in the sentences are automatically identified and transformed into a normal form consisting of two elements: the patient condition and the descriptive medical knowledge. In example 2, the sentence is transformed into the normal form shown in Table 3.

Table 3. Normal form of sentence in example 2

Patient Condition	All patients with symptomatic chronic heart failure that is caused by systolic left ventricular dysfunction
Descriptive Medical Knowledge	should receive an ACE-inhibitor

Patient Condition Representation. The patient condition is automatically represented by firstly identifying the central condition entities and then, the links among them, that can be *AND* or *OR* links. Following with example 2, the patient condition is represented by means of two concepts (*Chronic heart failure* with the modifier *symptomatic*, and *Left ventricular systolic dysfunction*) linked by the connective AND. Hence, patients with *Chronic heart failure* (modified by *symptomatic*) and *Left ventricular systolic dysfunction* verify 'They should receive an ACE-inhibitor'.

Table 4. Representation of a patient condition in example 2

Patient Condition	Representation
All patients with symptomatic chronic heart failure	((Disease or Syndrome) chronic heart failure (Functional concept) Symptomatic) AND
that is caused by systolic left ventricular dysfunction	((Pathologic Function) Left ventricular systolic dysfunction)

3.3   Relationship Extraction

The relationship extraction phase is carried out in six steps that are described below.

Identification of central information units. This step detects the central information units (i.e., leading concepts) in the sentence, in order to match them with a small set of predefined relationships. Entities corresponding to very generic terms, such as symptom, sign, finding, procedure, therapy, patient, etc. are considered non-leading concepts; whereas the rest of entities are considered leading concepts. In example 1, some concepts are classified as leading concepts (CHF, complete blood count, Electrolytes measurement, etc.) and others as non-leading concepts (patient, evaluation procedure, diagnosis). In example 2, there is only one leading concept (ACE-inhibitor), so the diagnosis context (in this case, CHF) is added to the set of leading concepts.

Generation of entity pairs. An entity pair is a pairing of entities recognized in the sentence, that may be the arguments of a relationship. All entities in the set of leading concepts are used to generate entity pairs with a constraint: entities that are paired must be in different external NPs. In example 1, six pairs are generated due to the leading concept *CHF* in the marked Pre-NP and the six leading concepts (*complete blood count, Electrolytes measurement*, etc.) in the marked Co-NP (See Table 1):

- Complete blood count↔Chronic heart failure
- Electrolytes measurement, serum↔Chronic heart failure
- Creatinine measurement↔Chronic heart failure
- Glucose measurement↔Chronic heart failure
- Measurement of liver enzyme↔Chronic heart failure
- Thyroid Function Tests↔Chronic heart failure

Import of relationships. This step assigns a relationship type to an entity pair. For this, it matches each generated entity pair with the relationships defined in the ontology, taking into account the semantic categories of each entity in the pair. In example 1, the imported relationship is 'diagnoses (T163)', a relationship between a Laboratory Procedure and a Pathologic Function; whereas in the example 2, there are two possible relationships in the ontology between a Pharmacologic Substance and a Pathologic Function: 'diagnoses (T163)' and 'treats (T154)'.

Disambiguation of relationships. In case of more than one imported relationship for a entity pair, non-leading concepts or keywords are used to disambiguate the relationships. In example 2, the keyword *receive* is used to discard the relationship 'diagnoses'.

Detection of negated relationships. Negated relationships are detected using regular expressions following a simplified version of the NegEx algorithm [27], based on triggering regular expressions to identify negations. This is the case in example 3.

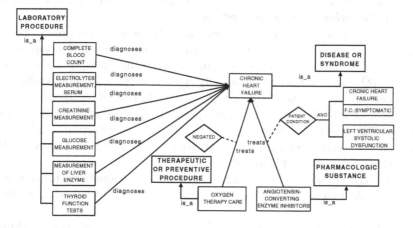

Fig. 2. Results of our method for the three running examples

Consistency checking. When a new relationship is imported a consistency checking is needed, in order to ensure that the new relationship is consistent with the previously imported relationships. The method checks whether the relationships is already found in the knowledge base or whether it is in conflict with another relationships already present in the knowledge base (such as one positive and another negative, one more specific than the other, etc).

Following the mentioned six steps, each sentence is mapped to an ontological representation of its relevant clinical content. In Fig. 2, we can see the results for the three running examples.

4   Results

A preliminary evaluation of the method performance was made in order to determine the method accuracy regarding mapping of natural language text to ontological structures. Ninety three sentences were underlined by the group of physicians taking part in the project. For each sentence, one or more relationships were generated using an implementation of our method. These automatically generated relationships are compared to relationships that were manually modeled by the authors. For the evaluation, the following criteria were considered. A relationship is interpreted as correct if it contains all diagnosis and treatment entities described in the sentence, and if the patient condition and the relationship represent the same knowledge as the manually created one.

Of the ninety three sentences, 81% were correctly and completely mapped to the corresponding Metathesaurus concepts and 65% were correctly represented by the relationships extracted by our method. The following errors led to incorrect relationships of sentences:

1. Erroneous syntactic analysis led to an incomplete entity recognition. OpenNLP failed to identify some nouns and it parses them as verbs, reducing the total of recognized entities. For example, the word *swelling* is parsed as a verb, so the term *ankle swelling* cannot be recognized.
2. Missing processing of some features modifying the procedures caused an incomplete representation of relationships. In most cases, the feature was modifying a non-leading concept. For example, in example 1 the feature *routine (Qualitative Concept)* is an adjective modifying the non-leading concept *evaluation,* so this feature plays no part in the relationships. Another specific case was the modifier *at rest (Functional Concept)*, which could not be recognized by our method. The modifier *at rest* is a prepositional phrases and our the method removes the preposition before requesting to the UMLS, so it cannot recognize this modifier.
3. Comparative syntactic structures resulted in a wrong identification of leading concepts. An example of incorrectly represented sentence is *anti-arrhythmic drugs other than beta-blockers.*
4. Complex syntactic sentences, including co-references and relative clauses, brought on a missing or wrong representation of patient conditions or

relationships. For example, the patient condition *patients in whom echocar-diography at rest has not provided enough information* is ignored by our method and therefore it is not represented.

5   Conclusions

The modeling process of CPGs is still a very complex and labor-intensive activity. In this paper, we propose an automated method to gradually map relevant clinical contents to an ontology, from natural language input. However, building a domain ontology is still a difficult task. Some current approaches [28] take an existing medical taxonomy such as Mesh as a starting point to build the domain ontology and then, this core ontology is extended with concepts from other domain resources. In this way, they may contribute enhancements to the core ontology. On the contrary, in our approach, we start from the UMLS resources and we identify guideline diagnosis and therapy entities (and relationships) of these large resources. Our approach is supported by the need for sharing a common understanding of the entities and relationships being used in the clinical domain.

This paper demonstrates the applicability of integrating several open-source tools and pre-existing algorithms, to parse guideline documents in order to identify firstly diagnosis and therapy entities and then, meaningful relationships among these entities. Our approach serves as an example and promising method to automatically encode diagnosis and treatment recommendations with mappings to standardized terminology. Our method offers a systematic approach to encode descriptive knowledge of a CPG in an ontology, using a standardized vocabulary. In particular, this information is relevant for decision support, promoting the guideline content to be delivered automatically at the point of care.

Acknowledgements. This work has been funded by the Ministerio de Educación y Ciencia, through the national research project HYGIA (TIN2006-15453-C04-02).

References

1. Rosser, W., Davis, D., Gilbart, E.: Promoting effective guideline use in ontario. JAMC 165(2), 181–182 (2001)
2. Grimshaw, J.M., Russel, I.T.: Implementing clinical practice guidelines: can guidelines be used to improve clinical practice? Effective Health Care 8, 1–12 (1994)
3. de Clercq, P.A., Blom, J.A., Korsten, H.H.M., Hasman, A.: Approaches for creating computer-interpretable guidelines that facilitate decision support. Artificial Intelligence in Medicine 31(1), 1–27 (2004); Review article
4. Isern, D., Moreno, A.: Computer-based execution of clinical guidelines: A review. International Journal of Medical Informatics 77(12), 787–808 (2008)

5. Miksch, S., Shahar, Y., Johnson, P.: Asbru: A task-specific, intention-based, and time-oriented language for representing skeletal plans. In: Motta, E., van Harmelen, F., Pierret-Golbreich, C., Filby, I., Wijngaards, N. (eds.) Proceedings of Seventh Workshop on Knowledge Engineering: Methods and Languages (KELM 1997), Milton Keynes, UK (1997)

6. Musen, M., Tu, S., Das, A., Shahar, Y.: EON: a component-based approach to automation of protocol-directed therapy. Journal of the American Medical Informatics Association 3, 367–388 (1996)

7. Boxwala, A., Peleg, M., Tu, S., Ogunyemi, O., Zeng, Q., Swang, D., Patel, V., Greenes, R., Shortlife, E.: GLIF3: a representation format for sharable computer-interpretable clinical practise guidelines. Journal of Biomedical Informatics 37, 147–161 (2004)

8. Ciccarese, P., Caffi, E., Quaglini, S., Stefanelli, M.: Architectures and tools for innovative health information systems: the guide project. International Journal of Medical Informatics 74, 553–562 (2005)

9. Carbonell, J., Etzioni, O., Gil, Y., Joseph, R., Knoblock, C., Minton, S., Veloso, M.: PRODIGY: an integrated architecture for planning and learning. ACM SIGART Bulletin 2(4), 51–55 (1991)

10. Sutton, D., Fox, J.: The syntax and semantics of the PROforma guideline modeling language. Journal of the American Medical Informatics Association 10, 433–443 (2003)

11. Sonnenberg, F., Hagerty, C.: Computer-interpretable clinical practice guidelines. where are we and where are we going? Yearb Med. Inform., 145–158 (2006); Review article

12. Kosara, R., Miksch, S., Shahar, Y., Johnson, P.: AsbruView: Capturing complex, time-oriented plans beyond ow-charts. Thinking with Diagrams, vol. 98, August 1998, pp. 22–23 (1998)

13. InferMed., Arezzo Technical White Paper, T.r., http://www.infermed.com/

14. Steele, R., Fox, J.: Tallis PROforma primer - introduction to proforma language and software with worked examples. Technical report, Advanced Computation Laboratory, Cancer Research, London, UK

15. Gennari, J., Musen, M., Fergerson, R., Grosso, W., Crubézy, M., Eriksson, H., Noy, N., Tu, S.: The evolution of Protègè: an environment for knowledge-based systems development. International Journal of Human-Computer Studies 58(1), 89–123 (2003)

16. Kaiser, K., Akkaya, C., Miksch, S.: How can information extraction ease formalizing treatment processes in clinical practice guidelines? a method and its evaluation. Artificial Intelligence in Medicine 39(2), 151–163 (2007)

17. Serban, R., ten Teije, A., van Harmelen, F., Marcos, M., Polo-Conde, C.: Extraction and use of linguistic patterns for modelling medical guidelines. Artificial Intelligence in Medicine 39(2), 137–149 (2007)

18. Rassinoux, A.M., Baud, R., Scherrer, J.R.: A multilingual analyser form medical texts (1994),
http://mbi.dkfz-heidelberg.de/helios/doc/nlp/Rassinoux94b.html

19. Friedman, C., Liu, H., Shagina, L., Johnson, S., Hripcsak, G.: Evaluating the UMLS as a source of lexical knowledge for medical language processing. In: Proc. AMIA Symp., pp. 189–193 (2001)

20. Denecke, K.: Semantic structuring of and information extraction from medical documents using the umls. Methods Inf. Med. 47(5), 425–434 (2008)

21. Bodenreider, O.: Lexical, terminological and ontological resources for biological text mining. In: Ananiadou, S., McNaught, J. (eds.) Text mining for biology and biomedicine, pp. 43–66. Artech House, Boston (2006)
22. Lindberg, D., Humphreys, B., Mc Cray, A.: The unified medical language system. Methods of Information in Medicine 32, 281–291 (1993)
23. Gruber, T.: Encyclopedia of Database Systems. In: Liu, L., Tamer, M. (eds.) Ontology
24. Gómez-Pérez, A., Fernández-López, M., Corcho, O.: Ontological Engineering: With Examples from the Areas of Knowledge Management, E-commerce and the Semantic Web. Springer, Heidelberg (2004)
25. Swedberg, K.: Guidelines for the diagnosis and treatment of chronic heart failure:executive summary (update 2005). European Heart Journal 26, 1115–1140 (2005)
26. Andreopoulos, B., Alexopoulou, D., Schroeder, M.: Word sense disambiguation in biomedical ontologies with term co-occurrence analysis and document clustering. Int. J. Data Mining and Bioinformatics 2(3), 193–215 (2008)
27. Chapman, W., Bridewell, W., Hanbury, P., Cooper, G.F., Buchanan, B.G.: A simple algorithm for identifying negated findings and diseases in discharge summaries. Journal of Biomedical Informatics 35(5), 301–310 (2001)
28. Diallo, G., Kostkova, P., Jawaheer, G., Jupp, S., Stevens, R.: Process of building a vocabulary for the infection domain. In: 21st IEEE International Symposium on Computer-Based Medical Systems, Jyvaskyla, Finland, June 17-19 (2008)

A Hybrid Methodology for Consumer-Oriented Healthcare Knowledge Acquisition

Elena Cardillo, Luciano Serafini, and Andrei Tamilin

FBK-IRST, Via Sommarive 18, 38100 Povo (TN), Italy
{cardillo,serafini,tamilin}@fbk.eu

Abstract. In Consumer Healthcare Informatics it is still difficult for laypersons to find, understand and act on health information, due to the persistent communication gap between specialized medical terminology and "lay" medical terminology used by healthcare consumers. Furthermore, existing clinically-oriented terminologies cannot provide sufficient support when integrated into consumer-oriented applications, so there is a need to create consumer-friendly terminologies reflecting the different ways healthcare consumers express and think about health topics. This work suggests a methodology to acquire consumer health terminology for creating a Consumer-oriented Medical Vocabulary for Italian that mitigates this gap. This resource, aligned to a standard medical terminology, could be useful in Personal Health Records to improve users' accessibility to their healthcare data. We performed evaluation mapping on acquired data to the International Classification of Primary Care (ICPC-2) to find overlaps and the candidate "lay" terms that can be considered good synonyms for the medical ones.

Keywords: Medical Classification Systems, Knowledge Acquisition, Consumer Healthcare Terminology, ICPC2 Mapping.

1 Introduction

With the advent of the Social Web and Healthcare Informatics technologies, we can recognize that a linguistic and semantic discrepancy still exists between specialized medical terminology used by healthcare providers or professionals, and the so called "lay" medical terminology used by patients and healthcare consumers in general. The medical communication gap became more evident when consumers started to play an active role in healthcare information access, becoming more responsible for their personal health care data, exploring health-related information sources on their own, consulting decision-support on the web, and using patient-oriented healthcare systems, which allow them to directly read and interpret clinical notes or test results and to fill in their Personal Health Record (PHRs). As a matter of fact, during this disintermediated interaction consumers can use only their own knowledge, experience and preferences, and this can often generate a wrong inference of the meaning of a term, or the mis-association of a term with its context [14].

Though much effort has been spent on the creation of medical resources (in particular thanks to the improvements in medical Knowledge Integration [9] facilitated by the use of Semantic Web Technologies [3]) used above all to help physicians in filling

D. Riaño et al. (Eds.): KR4HC 2009, LNAI 5943, pp. 38–49, 2010.

in Electronic Health Records (EHRs), facilitating the process of codification of symptoms, diagnoses and diseases (we can see an example in Duftschmid [5]), there is little work based on the use of consumer-oriented medical terminology, and in addition most existing studies have been done only for English. To help healthcare consumers fill this gap, the challenge is to sort out the different ways they communicate within distinct discourse groups and map the common, shared expressions and contexts to the more constrained, specialized language of healthcare professionals.

A consumer-oriented medical terminology can be defined as a *"collection of forms used in health-oriented communication for a particular task or need by a substantial percentage of consumers from a specific discourse group and the relationship of the forms to professional concepts"* [14] (e.g. *Nosebleed - Epistaxis*; *Hearth attack - Myocardial Infarction*; and other similar relations). It is used most of the time for three possible bridging roles between consumers and health applications or information: a) Information Retrieval, to facilitate automated mapping of consumer-entered queries to technical terms, producing better search results; b) Translation of Medical Records, supplementing medical jargon terms with consumers-understandable names to help patients interpretation; c) Health Care Applications, to help the integration of different medical terminologies and to provide automated mapping of consumer expressions to technical concepts (e.g. querying for the "lay" terms *Short of breath* and receiving information also for the corresponding technical concepts *Dyspnea*).

Figure 1 shows, for instance, how a consumer-oriented medical vocabulary can be useful providing translation functionalities, if integrated in healthcare systems, in two typical scenarios: 1) in the communication from professional to consumer, and 2) in the communication from consumer to professionals. Given this scenario, the present work proposes a hybrid methodology for the acquisition of consumer-oriented medical knowledge and "lay" terminology expressing particular medical concepts, such as symptoms and diseases, for the consequent creation of a Consumer-oriented Medical

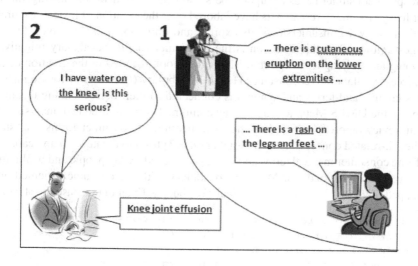

Fig. 1. Typical scenario for the use of a CHV

Vocabulary for Italian. We are particularly interested in performing analysis of the clinical mapping between this consumer-oriented terminology and the more technical one used in the International Classification of Primary Care (ICPC-2)[1], to find overlaps between them and to understand how many of these consumer-oriented terms can be used as good synonyms for the ICPC2 concepts. This consumer-oriented resource could be integrated with ICPC2 and other existing lexical and semantic medical resources, and used in healthcare systems, like PHRs, to help consumers during the process of querying and accessing healthcare information, so as to bridge the communication gap. The present work will be structured as follows: In Section 2 is described the State of the Art in the field of medical terminologies, both in clinically and consumer-oriented healthcare; in Section 3 we will present our approach, focusing on the Knowledge Acquisition Process (Section 3.1), the Terminology Extraction (Section 3.2) and the Mapping Analysis (Section 3.3); in Section 4 are presented preliminary results; and finally in Section 5 are proposed concluding remarks and some future works.

2   Consumer-Oriented Medical Terminologies

Over the last two decades research on Medical Terminologies has become a popular topic and the standardization efforts have established a number of terminologies and classification systems as well as conversion mappings between them to help medical professionals in managing and codifying their patients health care data, such as UMLS Metathesaurus[2], SNOMED International[3], ICD-10[4] and the already mentioned ICPC-2. They concern, in fact, *"the meaning, expression, and use of concepts in statements in the medical records or other clinical information systems"* [11]. Despite of the wide use of these terminologies, the vocabulary problem continues to plague health professionals and their information systems, but also consumers and in particular laypersons, who are the most damaged by the increased communication gap.

To respond consumer needs to support personal healthcare decision-making, during the last few years, many researchers have labored over the creation of lexical resources that reflect the way consumers/patients express and think about health topics. One of the largest initiatives in this direction is the Consumer Health Vocabulary Initiative[5], by Q. Zeng and colleagues at Harvard Medical School, resulted in the creation of the Open Access Collaborative Consumer Health Vocabulary (OAC CHV) for English. It includes lay medical terms and synonyms connected to their corresponding technical concepts in the UMLS Metathesaurus. They combined corpus-based text analysis with a human review approach, including the identification of consumer forms for "standard" health-related concepts. Also Soergel *et al.* [13] tried to create such a vocabulary identifying consumer medical terms and expressions used by lay people and health mediators, associating a Mediator Medical Vocabulary with the consumer-oriented one, and mapped them to a Professional Medical Vocabulary. Even in this case the standard

---

[1] http://www.globalfamilydoctor.com/wicc/icpcstory.html
[2] http://www.nlm.nih.gov/research/umls/
[3] http://www.ihtsdo.org/snomed-ct/
[4] http://www.who.int/classifications/icd/en/
[5] http://www.consumerhealthvocab.org

terminology used for mapping was UMLS. These and other similar studies examined large numbers of consumer utterances and consistently found that between 20% and 50% of consumer health expressions were not represented by professional health vocabularies. Furthermore, a subset of these unrepresented expressions underwent human review. In most of these cases they performed automatic term extraction from written texts, such as healthcare consumer queries on medical web sites, postings and medical publications. An overview of all these studies can be found in Keselman *et al.* [6].

It is important to stress that there are only few examples of the real application of the most of initiatives. For example, in Kim *et al.* [7] and Zeng *et al.* [15] we find an attempt to face syntactic and semantic issues in the effort to improve PHRs readability, using the CHV to map content in EHRs and PHRs. On the other hand, Rosembloom *et al.* [12] developed a clinical interface terminology, a systematic collection of healthcare-related phrases (terms) to support clinicians' entries of patient-related information into computer programs such as clinical "note capture" and decision support tools, facilitating display of computer-stored patient information to clinician-users as simple human-readable texts. Concerning multilingual consumer-oriented health vocabularies, we can only mention the initiative of the European Commission Multilingual Glossary of Popular and Technical Medical Terms[6], in nine European languages, but it is a limited medical vocabulary for medicinal product package inserts accessible to consumers. In fact, it consists of a list of 1,400 technical terms frequently encountered in inserts, with corresponding consumer terms in all the languages of the EC. Greater overlap between technical and lay terms was observed for the Romance languages and Greek than for the Germanic languages (except English) and some technical terms had no lay equivalent.

3 Approach

In this study we focused on a hybrid methodology for the acquisition of consumer-oriented knowledge (lay terms, words, and expressions) used by Italian speakers to identify "symptoms", "diseases", and "anatomical concepts". Three different target groups were considered for the application of our approach: First Aid patients subjected to a Triage Process; a community of Researchers and PhD students with a good level of healthcare literacy; and finally a group of elderly people with a modest background and low level of healthcare literacy. The proposed methodology consists of the following steps:

1. Familiarization with the domain and exploitation of existing common lexical resources (Glossaries, Thesauri, Medical Encyclopedias, etc.);
2. Application of three different Elicitation Techniques to each group;
3. Automatic Term Extraction and analysis of acquired knowledge by means of a Text Processing tool;
4. Clinical review of extracted terms and manual mapping to a standard medical terminology (ICPC2), performed by physicians;
5. Evaluation of results in order to find candidate terms to be included in the Consumer-oriented Medical Vocabulary.

---

[6] http://users.ugent.be/~rvdstich/eugloss/information.html

3.1   Knowledge Acquisition Process

The Knowledge Acquisition process has the aim of identifying and capturing knowledge assets and terminology to populate a knowledge repository for a specific domain. A major part of this process is capturing knowledge from experts (who in our case correspond to laypersons and common people due to the type of medical knowledge/terminology we want to extract), a task that is made cost-effective and efficient by using knowledge models and special elicitation techniques. These techniques should be used in different phases of the process, since each of them permits capturing a specific typology of knowledge and achieving specific aims. An overview of the most common elicitation techniques can be found in Milton [8]. In this study, we applied three different methods for acquiring "lay" medical knowledge/terminology, from the already mentioned groups of people: 1) Collaborative Wiki-based Acquisition; 2) Nurse-assisted Acquisition; and 3) Interactive Acquisition combining traditional elicitation techniques (*Focus Groups, Concept Sorting and Games*). These can be described as follows:

Wiki-Based Acquisition. The first method of acquisition is based on the use of a Semantic Media Wiki system[7], an easy to use collaborative tool, allowing users to create and link, in a structured and collaborative manner, Wiki pages on a certain domain of knowledge. Using our on-line *eHealthWiki* system[8], users created Wiki pages for describing symptoms and diseases, using "lay" terminology, specifying in particular the corresponding anatomical categorization, the definition and possible synonyms. The system has been evaluated over a sample of 32 people: researchers, PhD students and administrative staff of our research institute (18 females, 14 males, between 25 and 56 years old). Figure 2 shows an example of wiki page in which users described the symptom "Absence of Voice" *Abbassamento della voce*, providing definition in lay terms, anatomical localization, synonyms.

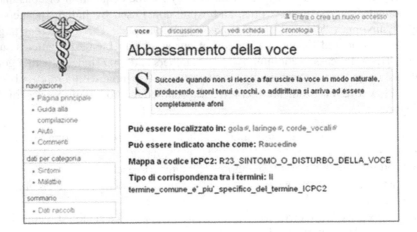

Fig. 2. Wiki page example created by users to express the symptom Absence of voice

---

[7] http://semantic-mediawiki.org/wiki/Semantic-MediaWiki
[8] http://ehealthwiki.fbk.eu

In one month, we collected 225 Wiki pages, 106 for symptoms and 119 for diseases, and a total of 139 synonyms for the inserted terms. During this test users were reluctant to use the collaborative functionality of the Wiki system, which permits modification of concepts added by others, even in the case of evident mistakes in term definitions or categorization. Some examples of categorization mistakes that had not been modified are "Singhiozzo" (*Hiccup*), and "Mal di Testa"(*Headache*), both categorized as disease instead of symptom. These and others similar incongruences highlighted the fact that users had problems in categorizing medical terms - mainly due to their clinic ambiguity - and also their erroneous daily use of these terms.

Nurse-assisted Acquisition. The second technique involved nurses of a First Aid Unit in a Hospital of the Province of Trento[9] as a figure of mediation for the acquisition of terminology about patient symptoms and complaints, helping them to express their problems using the classical subjective examination performed during the Triage Process[10]. This acquisition method involved 10 nurses, around 60 patients per day and a total of 2.000 Triage Records registered in one month. During this period nurses acquired the principal problems expressed by their patients using "lay" terminology and inserted them into the Triage Record together with the corresponding medical concepts usually used for codifying patient data (i.e. the expression "Ho i crampi alla pancia" - *I have a stomach ache* - inserted together with the corresponding medical concept "Addominalgia" - *Abdominal pain*).

Focus-Group Acquisition. The last method used in our study consisted in merging three different traditional elicitation techniques: Focus Group, Concept Sorting, and Board Games, in order to allow interaction and sharing circumstances to improve the process of acquisition. The target in this case was a community of 32 elderly people in a Seniors Club, aged from 65 to 83 year old. We used groups activities (four groups divided according to a specific body part category, i.e. head and neck, abdomen and back, arms and chest, pelvic area and legs) to acquire lay terms and expressions for symptoms, diseases and anatomical concepts. They were asked to write on little cards all known symptoms and diseases related to the assigned area, comparing their idea with other members of the group to find a common definition for the written terms. About 160 medical terms were collected, which, at the end of the process, were analyzed together with other groups, creating discussions, exchanging opinions on terms definitions, synonyms, and recording preferences and shared knowledge. In particular, all participants gave preferences for choosing the right body system categorization (digestive, neurological, respiratory, endocrine, etc.) of each of the written concepts. This allowed us not only to extract lay terminology, but also to understand how elderly people define and categorize medical concepts, in order to compare these results with that obtained from the other two mentioned techniques. To give an example of the acquisition process, elderly people in the second group, responsible for the collection of terms related to the body area "abdomen and back", collected lay terms such as *fuoco di Sant'Antonio* (Shingles) or *sfogo* (Rash) corresponding to the medical term "Herpes

---

[9] http://www.apss.tn.it/Public/ddw.aspx?n=26808

[10] The Triage activity has the aim to prioritize, by means of a few minutes examinations, patients based on the severity of their condition.

Zoster", describing it as *a rash on the lower extremity of the back, due to an allergic reaction to drugs, food or to the contact with plants*, and finally they categorized it as a medical concept belonging to the "integument system".

3.2   Term Extraction

Three sets of collected data were further processed and analyzed, to detect candidate consumer-oriented terms, by means of Text-2-Knowledge (T2K) tool, developed at the Institute of Computational Linguistics of Pisa (Italy)[11]. This tool allowed us to automatically extract terminology from the data sets derived from the KA process and to perform typical text processing techniques (normalization, POS tagging, chunking, etc), but also to calculate statistics on the extracted data such as term frequency. The computational analysis system adopted by the tool includes a specific plug-in for the analysis of Italian. It provides, as final output, a term-based vocabulary whose added value is represented by the terms' semantic and conceptual information regarding the vocabulary itself. These terms, which can be either single or multi-word terms, are organized in a hierarchical hyponym/hyperonym relation depending on the internal linguistic structure of the terms [1]; that is, by sharing the same lexical head.

In spite of the advantages of the automatic extraction process, allowing for extraction of many compound terms, such a procedure has demonstrated that a large number of terms, certainly representative of consumer medical terminology, were not automatically extracted, since, due to the quantitative limits of the corpus dimensions, their occurrence was inferior with respect to the predefined threshold value. Consequently, we performed an additional manual extraction to take into account such rare terms, usually mentioned by a single participant. Statistical results for the three different data sets are further discussed in Section 4.

3.3   Clinical Review and Mapping Analysis

Terms extracted by T2K were further reviewed by two physicians to find mistakes and incongruities in categorization and synonymy. In particular, many mistakes were found by physicians in the first set of terms (Wiki-based), where a wrong categorization was assigned to 25 terms, and where wrong synonyms were expressed for 8 terms. They found similar incongruities in the third set (Elderly people), where wrong categorizations were assigned to 40 terms, e.g. "Giramento di Testa" or "Vertigini" (*Vertigo* or *Dizziness*), categorized in the Cardiovascular System instead of the right Neurological one. Concerning the second data set, clinical review was not performed, because it was directly performed firstly by a nurse and then by a physician during the process of Triage.

During the second part of our clinical review physicians were asked to map a term/medical concept pair by using a professional health classification system, the above mentioned International Classification for Primary Care 2nd Edition (ICPC2-E, electronic version) [10], which has received great widespread and preference within the European Union. It addresses fundamental parts of the healthcare process: it is used

---

[11] http://www.ilc.cnr.it

in particular by general practitioners for encoding symptoms and diagnoses. It has a biaxial structure that consider medical concepts, related to *symptoms, diseases and diagnoses*, and *medical procedures*, according to 17 Problem Areas/Body Systems. In a previous work we encoded ICPC-2-E using the recently developed Web Ontology Language (OWL) [2] (both for English and Italian), and we also provided the formalization of the existing clinical mapping with the ICD10 classification system, as shown in Cardillo *et al.* [4].

By means of the mapping between "lay" terms and ICPC2 concepts we want to reconstruct the meaning (concept) inherent in the lay usage of a term, and then to agree that consonance between lay and professional terms exists on the basis of this deeper meaning, rather than the lexical form. We identified five different types of relations between consumer terms and ICPC2 medical concepts:

- Exact Mapping between the pairs; this occurs when the term used by a layperson can be found in ICPC2 rubrics and both terms correspond to the same concept. For instance, the lay term "Febbre" (*Fever*) would map to the ICPC2 "Febbre"(*Fever*) term, and both will be rooted to the same concept.
- Related Mapping; this involves lay synonyms and occurs when the lay term does not exist in the professional terminology, but corresponds to a professional term that denotes the same (or closely related) concept. For instance, the lay term "Sangue dal Naso" (*Nosebleed*) corresponds to "Epistassi" (*Epistaxis*) in ICPC2.
- Hyponymy Relation; this occurs when a lay term can be considered as a term of inclusion of a ICPC2 concept. For instance, the lay term "Abbassamento della Voce" (*Absence of Voice*) is included in the more general ICPC2 concept "Sintomo o disturbo della voce" (*Voice Symptom/Complaint*).
- Hyperonymy Relation; in this case the lay term is more general than one or more ICPC2 concepts, so it can be considered as its/their hyperonym. For example, the term "Bronchite" (*Bronchitis*) is broader than "Bronchite Acuta/Bronchiolite" (*Acute Bronchitis/Bronchiolitis*) and "Bronchite Cronica" (*Chronic Bronchitis*) ICPC2 concepts.
- Not Mapped; those lay terms that cannot be mapped to the professional terminology. These can be legitimate health terms, the omission of which reflects real gaps in existing professional terminologies; or they can represent unique concepts reflecting lay models of health and disease. For example, the lay term "Mal di mare" (*Seasickness*).

4 Evaluation

As previously mentioned, our methodology of acquisition allowed us to acquire varied consumer-oriented terminology and to perform an interesting terminological and conceptual analysis. Tables 1-4 provide term extraction and mapping evaluation in terms of a statistical analysis. By means of the term extraction process, we were able to extract a total of 962 medical terms from 225 Wiki pages, 375 of which were not considered pertinent to our aim. We performed mapping analysis on 587 terms (306 symptoms, 140 diseases, and 141 anatomical concepts) as summarized in Table 1.

Table 1. Wiki term collection

	Tot. Terms	Exact Map.	Related Map.	Hyponyms.	Hyperonyms
Symptoms	306	26	50	40	9
Diseases	140	42	19	38	38
Anatomy	141	105	11	16	4
Other	375	/	/	/	/
Tot.	962				
Not Mapped:	186				

We can observe that most of the exact mappings with ICPC2 are related to anatomical concepts, and that many synonyms in lay terminology and inclusion terms were found for symptoms. Table 2 shows the results related to the mapping analysis for the data extracted by Nurse-assisted acquisition.

Table 2. Nurse-assisted term collection

	Tot. Terms	Exact Map.	Related Map.	Not Mapped
Symptoms	508	134	197	177
Diseases	325	86	94	145
Anatomy	275	120	95	60
Other	1281	/	/	/
Tot.	2389			

We extracted a total of 2389 terms from 2000 Triage records, but about half of these terms were considered irrelevant for our evaluation, so mapping was provided only for 1108 terms (508 symptoms, 325 diseases, and 275 anatomical concepts) as shown in the table. Contrary to the previous results, here is interesting to highlight the high presence of lay terms used for expressing symptoms with exact mappings to ICPC2, but also many synonyms in lay terminology for ICPC2 symptoms and diseases. This is particularly related to the context chosen for the acquisition, where patients just ask for help about suspected symptoms and complaints. Concerning the last data set, 321 medical terms were extracted by the transcription of the Focus Group/Game activity, but just 243 (79 symptoms, 87 diseases, and 77 anatomical concepts) were relevant for mapping analysis. As shown in Table 3, here all the symptoms extracted have corresponding medical concept in ICPC2 terminology.

Table 3. Focus Group /Game with Elderly Persons

	Tot. Terms	Exact Map.	Related Map.	Not Mapped
Symptoms	79	35	44	0
Diseases	87	29	54	4
Anatomy	77	51	18	8
Other	78	/	/	/
Tot.	321			

Finally, Table 4 compares the three data sets together and shows that the most profitable method for acquiring consumer-oriented medical terminology was the one assisted by Nurses. But the limit of this method is that it is time-consuming for nurses who have to report all the patient "lay" health expressions. Concerning Wiki-based method, even if not exploited for the collaborative characteristic, has demonstrated good qualitative and quantitative results. Furthermore, can be considered interesting the results concerning mapping to ICPC2, because 2/3 of the terms extracted are covered by ICPC2 terminology.

Table 4. Results Overview

Sources	Tot. Terms	Tot. Mapped	Not Mapped
Wiki-based collection	962	398	186
Nurse-assisted collection	2389	726	382
Focus Groups collection	321	231	12
Tot.	3662	1355	580

To conclude our evaluation we have to highlight that comparing the three sets of extracted terms, the overlap is only of 60 relevant consumer medical terms. The total overlap with ICPC2 is about of 508 medical concepts on a total of 706 ICPC2 concepts. This means that all the other mapped terms can be considered synonyms or quasi synonyms of the ICPC2 concepts. The large number of Not Mapped terms and the low overlap between the three sets of extracted terms demonstrate that we extracted a very variegated range of medical terms, many compound terms and expressions, which can be representative of the corresponding technical terms present in standard medical terminologies, and which can be used as candidate for the construction of our Consumer-oriented Medical Vocabulary for Italian.

5   Conclusion and Future Work

In this paper we have presented a hybrid methodology for acquiring consumer-oriented medical Knowledge and Terminology for Italian, consisted of lay expressions and terms used to indicate *symptoms*, *diseases* and *anatomical concepts*. We applied three exploratory elicitation techniques to three different samples of people, and we compared results on the basis of a term extraction process, for statistical analysis, and on a clinical mapping procedure, for finding overlaps between extracted lay terms and specialized medical concepts in the ICPC2 terminology. Comparing our approach with that followed by other researchers mentioned in Section 2, who developed consumer-oriented health vocabularies working only on big written corpora (forum postings and queries to medical websites), using machine learning algorithm and statistical methods (naive Bayesian classifiers, C-value, etc.) to extract consumer-oriented terminology, we gave more importance to qualitative data, focusing on different methods for acquiring medical lay terminology and knowledge directly from consumers in different scenarios related to General Practice. This allowed us not only to acquire data but also to try to understand how consumers make good or wrong use of medical terminology, how common expressions daily used in health communication really match to medical concepts used by professionals.

In practical terms, our methodology showed encouraging results because it allowed us to acquire many consumer-oriented terms, a low overlap with ICPC2 medical concepts, and a high number of related mappings (most of the time synonyms) to the referent medical terminology. Taking each of these acquisition techniques alone, we have to admit that one of their limits is that they do not allow to extract lay terminology with a good coverage of the whole domain of pathology and symptomatology. But using a hybrid approach in merging these techniques, and involving a more varied sample of people would improve the results, both from the qualitative and the quantitative point of view. Another limit could be seen in the process of manual mapping performed by physicians. After this pilot study we plan to perform a semi-automatic procedure for mapping lay and specialized terminology, which will be associated to the process of automatic term extraction, and validated by the review of physicians.

To improve the results of the Knowledge Acquisition process and to extract more variegated consumer-oriented terminology, not related to the regional context, we are analyzing written corpora, which include forum postings of an Italian medical website for asking questions to on-line doctors [12]. This would allow to extend our sample, covering a wider range of ages, people with different background and consequently different levels of healthcare literacy. This task will be very interesting for comparing results with that came out from the previous elicitation methods, both in quantitative and qualitative terms. Data extracted in this way will be used to validate the acquired terminology, by providing preferences between terms according to frequency and familiarity score.

References

1. Bartolini, R., Lenci, A., Marchi, S., Montemagni, S., Pirrelli, V.: Text-2-knowledge: Acquisizione semi-automatica di ontologie per l'indicizzazione semantica di documenti. Technical Report for the PEKITA Project, ILC, Pisa, p. 23 (2005)
2. Bechhofer, S., Van Harmelen, F., Hendler, J., Horrocks, I., McGuinness, D.L., Patel-Schneider, P.F., Stein, A.L.: OWL Web Ontology Language Reference, W3C Recommendation (2004)
3. Berners-Lee, T., Hendler, J., Lassila, O.: The Semantic Web. Scientific American 284(5), 34–43 (2001)
4. Cardillo, E., Eccher, C., Tamilin, A., Serafini, L.: Logical Analysis of Mappings between Medical Classification Systems. In: Dochev, D., Pistore, M., Traverso, P. (eds.) AIMSA 2008. LNCS (LNAI), vol. 5253, pp. 311–321. Springer, Heidelberg (2008)
5. Grabenweger, J., Duftschmid, G.: Ontologies and their Application in Electronic Health Records. In: eHealth2008 Medical Informatics meets eHealth, Wien, May 29-30 (2008)
6. Keselman, A., Logan, R., Smith, C.A., Leroy, G., Zeng, Q.: Developing Informatics Tools and Strategies for Consumer-centered Health Communication. Journal of Am. Med. Inf. Assoc. 14(4), 473–483 (2008)

---

[12] http://medicitalia.it

7. Kim, H., Zeng, Q., Goryachev, S., Keselman, A., Slaughter, L., Smith, C.A.: Text Characteristics of Clinical Reports and Their Implications for the Readability of Personal Health Records. In: 12th World Congress on Health (Medical) Informatics, MEDINFO 2007, pp. 1117–1121. IOS Press, Amsterdam (2007)

8. Milton, N.R.: Knowledge Acquisition in Practice: A Step-by-step Guide. Springer, Berlin (2007)

9. Nardon, F.B., Moura, L.A.: Knowledge Sharing and Information Integration in Healthcare using Ontologies and Deductive Databases. In: MEDINFO 2004, pp. 62–66. IOS Press, Amsterdam (2004)

10. Okkes, I.M., Jamoullea, M., Lamberts, H., Bentzen, N.: ICPC-2-E: the electronic version of ICPC-2. Differences from the printed version and the consequences. Family Practice 17, 101–107 (2000)

11. Rector, A.: Clinical Terminology: Why is it so hard? Methods of Information in Medicine 38(4), 239–252 (1999)

12. Rosembloom, T.S., Miller, R.A., Johnson, K.B., Elkin, P.L., Brown, H.S.: Interface Terminologies: Facilitating Direct Entry of Clinical Data into Electronic Health Record Systems. Journal of Am. Med. Inf. Assoc. 13(3), 277–287 (2006)

13. Soergel, D., Tse, T., Slaughter, L.: Helping Healthcare Consumers Understand: An "Interpretative Layer" for Finding and Making Sense of Medical Information. In: International Medical Informatics Association's Conference, IMIA 2004, pp. 931–935 (2004)

14. Zeng, Q., Tse, T.: Exploring and Developing Consumer Health Vocabularies. J. of Am. Med. Inf. Assoc. 13, 24–29 (2006)

15. Zeng, Q., Goryachev, S., Keselman, A., Rosendale, D.: Making Text in Electronic Health Records Comprehensible to Consumers: A Prototype Translator. In: 31st American Medical Informatics Association's Annual Symposium, AMIA 2007, pp. 846–850 (2007)

Identifying Disease-Centric Subdomains in Very Large Medical Ontologies: A Case-Study on Breast Cancer Concepts in SNOMED CT. Or: Finding 2500 Out of 300.000

Krystyna Milian[1], Zharko Aleksovski[2], Richard Vdovjak[2], Annette ten Teije[1], and Frank van Harmelen[1]

[1] Vrije Universiteit Amsterdam
krystyna.milian@few.vu.nl
[2] Philips Research
zharko.aleksovski@philips.com

Abstract. Modern medical vocabularies can contain up to hundreds of thousands of concepts. In any particular use-case only a small fraction of these will be needed. In this paper we first define two notions of a disease-centric subdomain of a large ontology. We then explore two methods for identifying disease-centric subdomains of such large medical vocabularies. The first method is based on lexically querying the ontology with an iteratively extended set of seed queries. The second method is based on manual mapping between concepts from a medical guideline document and ontology concepts. Both methods include concept-expansion over subsumption and equality relations. We use both methods to determine a breast-cancer-centric subdomain of the SNOMED CT ontology. Our experiments show that the two methods produce a considerable overlap, but they also yield a large degree of complementarity, with interesting differences between the sets of concepts that they return. Analysis of the results reveals strengths and weaknesses of the different methods.

Keywords: identifying ontology subdomain, disease related concepts, ontology subsetting, mapping medical terminologies, seed queries, medical guidelines.

1 Introduction

Large medical ontologies such as SNOMED CT [1] contain hundreds of thousands of clinical concepts usually organized in a hierarchy and interconnected by domain specific relations, together representing the explicit semantic knowledge describing a medical field. Such knowledge can be of great help when developing intelligent clinical decision support systems that focus on reasoning about patient data within a certain disease domain. A disease-specific, richly annotated

---

[1] http://www.ihtsdo.org/snomed-ct/

D. Riaño et al. (Eds.): KR4HC 2009, LNAI 5943, pp. 50–63, 2010.

semantic subdomain is also an important element in the process of overcoming the frequent problem of lexical heterogeneity between the concepts occurring in the patient data and those from the applicable clinical guidelines. However, identifying a *disease-centric subdomain* of a large medical ontology is not a trivial task. The relevant concepts are seldom to be found under one sub-branch of the ontology, instead they are usually scattered in various branches representing different facets of the domain coverage, e.g. clinical findings, procedures, anatomic regions, etc.

In this paper we describe a study on the identification of SNOMED CT concepts related to breast cancer. We compare results of two different methods: *(i)* The *seed query method* from [1] was used for extraction of concepts that are unique to breast cancer. *(ii)* The so-called *guideline-based method*, consisting of a manual mapping between SNOMED CT concepts and the important concepts from the Dutch national breast cancer guidelines, was used for the identification of those concepts that are relevant with respect to breast cancer.

Our experiments show that the two methods produce a considerable overlap, but they also yield a large degree of complementarity, with interesting differences between the sets of concepts that they return. The size of the identified subdomains is considerably smaller than that of the whole medical ontology (between 0.1%-1%), making the reasoning as well as the maintenance task of such subdomain much more feasible.

The paper is structured as follows: Section 2 introduces different notions of relevancy in subdomains of a medical ontology, and puts forward the main hypothesis of the paper. Section 3 and 4 introduce our two different subdomain-selection methods: the seed query method in section 3 and the guideline-based method in section 4. Section 5 compares and analyses the results. Section 6 presents related work. Section 7 summarizes the findings and presents the concluding remarks.

2   Two Types of Disease-Centric Subdomains

Before investigating methods for identifying disease-centric subdomains from a large ontology, we must first define what we mean by such a subdomain. For the purpose of this paper we will set our own definitions. Presented below methods are not based on any *a priori* modularization of the ontology, but they identify subdomains that are specific for any particular use of a vocabulary.

Definitions: We distinguish two kinds of disease-centric subdomains, namely *relevant subdomains* and *key subdomains*, which consist of relevant concepts and key concepts respectively. The notions of "relevant concepts" and "key concepts" are each defined as follows:

Relevant concept: A concept C is a *relevant concept* for a disease D if clinical guidelines for D state that it influences a decision on the diagnosis or treatment of D.

An example of a concept that is relevant to breast cancer is "pregnancy": datasources about breast cancer (such as guidelines, patient-records, textbooks, etc) often contain the concept "pregnancy" because certain treatments (e.g. chemotherapies) are ruled out for pregnant women.

However, the converse is not the case: not any document containing the concept "pregnancy" is likely to be about breast cancer. To capture this, we define a second notion:

Key concept: A concept C is a *key concept*[2] for a disease D if the occurrence of C in a datasource S means that S is conclusively about D.

An example is the concept "malignant neoplasm of breast". Any key concept is of course a relevant concept, but not vice versa.

Hypothesis: Our hypothesis is that the seed query method (described in section 3), when seeded properly, will identify only key concepts, while the manual guideline-based method (described in section 4) will identify relevant concepts. From the above definitions, this hypothesis also implies that the seed query results should be contained in the guideline-based results.

Choice of dataset: In this paper, we focus on breast cancer as our clinical domain both because of its prevalence and the highly progressed state-of-the-art in diagnoses and treatment, which is expected to involve a relatively rich vocabulary and thus presents an interesting use-case. We concentrate on SNOMED CT as our main ontology, mainly because of its high adoption and a broad clinical coverage, containing more than 300.000 concepts. Besides applying both methods to the breast cancer domain in SNOMED CT, we also apply the seed query method to three other very large ontologies, namely to NCI, MeSH and ICD10. We do this to verify the consistency of our results. The precise use of these ontologies is described in next section. Also applying the manual guideline-method to these ontologies would have been prohibitively expensive.

3   Seed Query Method to Find Key Concepts

Method. The seed query method, originally published in [1], is a combination of a lexical and a structural approach.

It takes a list of combinations of some of key concepts (the so-called "seed queries"), which serve as prior knowledge, to find an initial set of domain specific, in this case breast cancer related concepts through lexical mapping to the concepts in the ontology. This set is then expanded through the hierarchical structure of the ontology, and through the structure of UMLS (Unified Medical Language System[3]) metathesaurus. Given a set of seed queries, the process is completely automatic, ensuring repeatability of the extraction. It also allows for gradual improvement by adjusting the initial set of seed queries.

---

[2] "key" is inspired by the database notion of the same name
[3] http://www.nlm.nih.gov/research/umls/

In more detail, the seed query method proceeds in three steps: *(i)* *Query matching* which uses the concept's names, *(ii)* *Subconcept expansion* based on the hierarchical structure of the ontologies, and *(iii)* UMLS expansion which uses the UMLS metathesaurus. The three steps in this method are sequential, increasing the set incrementally, each step produces set of concepts which is passed as input to the next step. The third step produces the final result of the method. Next, we elaborate each of the steps, and also present it as a pseudo-code algorithm.

Query matching uses a list of seed queries to find concepts from the subdomain by trying to lexically match the queries to each concept from the ontology. The lexical match was not sensitive to letter capitalization, and in addition, Porter's stemmer algorithm [11] was used to normalize the words before comparison. Such queries consist of keywords or combinations of keywords which are specific to the subdomain, and when a concept lexically matches to some of these queries, it can be considered part of the subdomain. The algorithm for query matching is shown in Figure 1. It is applied on each of the four candidate ontologies separately.

Subconcept expansion expands the set of concepts produced in the first step by including their subconcepts. Each ontology generally organizes the concepts in a hierarchy through IS-A relations among them (e.g. Breast cancer IS-A Cancer). These relations were used to find all the subconcepts of the concepts found in the first step. This process was done exhaustively, transitively adding the subconcepts of the newly found concepts as well, until no new concepts could be added. The algorithm for subconcept expansion is shown in Figure 2. It is separately applied on each set obtained in the first step.

UMLS *expansion* uses UMLS to further increase the set produced in the second step. UMLS assigns a unique identifier to every concept from every ontology integrated in it, and if two concepts have the same identifier then they mean the same thing. Suppose two arbitrary concepts A_1 and A_2 from two ontologies ONT_1 and ONT_2 respectively, are assigned the same identifier in UMLS. Now, if A_1 is found as key concept in the first two steps for the ontology ONT_1 and A_2 is *not* found as key concept for the ontology ONT_2 in the first two steps, then A_2 can be added as a key concept for the ontology ONT_2, thus expanding the set of key concepts for the ontology ONT_2. This way of expanding the sets of key concepts is the third step of the method. It is done exhaustively, for every concept and every pair of ontologies used in the experiment. The algorithm for UMLS expansion is given in Figure 3. It is applied on the four sets of key concepts obtained in the second step.

Results. The breast cancer-centric subdomain of SNOMED CT (containing only key concepts for breast cancer) was extracted using the method described above.

We seeded the method with a hand-crafted list of breast cancer seed queries, shown in Table 1. After starting with a small number of key concepts, and iteratively adding seeds, we observed that after a small number of concepts the results stabilise, and no longer grow when adding further key concepts as seeds. This process has up to now been informal, and would merit a more detailed study in its own right.

The resulting set of matched concepts is empty in the beginning
1   subdomain := ∅
Lexically matching the concepts from the ontology to the query list
2   **for** each query $Q \in$ list of queries **do**
3   **for** each concept $C \in C^{ONT}$ **do**
4   **if** LEXICALMATCH(C, Q) **and** $C \notin$ subdomain **then**
5   subdomain := subdomain $\cup \{C\}$

Fig. 1. Step one: Query matching

Add all the subconcepts to the concepts in subdomain
1   **while** adding new concepts in subdomain is possible **repeat**
2   **for** each concept $X \in$ subdomain **do**
3   **for** each concept $Y \in C^{ONT}$ **do**
4   **if** $Y \subseteq X$ **and** $Y \notin$ subdomain **then**
5   subdomain := subdomain $\cup \{Y\}$

Fig. 2. Step two: subconcept-based expansion

Expanding each of 4 result sets through UMLS
1   **for** any $ONT_p, ONT_q \in \{$SNOMED CT, NCI, MeSH, ICD10$\}$, where $p \neq q$ **do**
2   **for** each concept $X \in$ subdomain$_p$ **do**
3   **for** each concept $Y \in$ subdomain$_q$ **do**
4   **if** UMLS $: X \equiv Y$ **and** $Y \notin$ subdomain$_p$ **then**
5   subdomain$_p$:= subdomain$_p$ $\cup \{Y\}$

Fig. 3. Step three: UMLS-based expansion

Besides SNOMED CT, the method was applied on three other ontologies: NCI[4] - a vocabulary for annotating medical documents primarily cancer related, MeSH[5] - a vocabulary for scientific literature annotation and ICD10[6] - a classification of diseases. The ontologies were used as extracted from the UMLS 2008AA version.

The results of applying the seed query method are shown in Table 2. The table shows that only a fraction of the entire ontology (much less than 1%) are key concepts for a disease such as breast cancer. It also shows that most of the results are actually found in the first phase. This is reasonable: most of the concepts are very specialized and are hence leafs in the ontologies. Finally, it is interesting to see that the most specialised ontology (the oncology-specific NCI) has the highest hit-rate of key concepts, and the most general and wide ranging ontologies (MeSH and SNOMED CT) have the lowest hit-rates.

---

[4] http://nciterms.nci.nih.gov
[5] http://www.nlm.nih.gov/mesh
[6] http://www.ahima.org/icd10

Table 1. Seed queries used to extract the breast cancer subdomain

1. Breast cancer
2. Breast carcinoma
3. Microcalcification
4. Mammary carcinoma
5. Lobular carcinoma
6. Ductal carcinoma
7. Mastectomy
8. Paget breast
9. HER2/neu
10. HER-2
11. BRCA

Table 2. Results of applying the seed query method on the four ontologies: incremental results are reported after each step (full method = after step 3)

Ontology	size of ontology	number of concepts extracted			% of full ontology
		after step 1	after step 2	after step 3	
SNOMED	308,677	198	271	279	0.09%
NCI	62,969	358	388	399	0.63%
MeSH	282,425	105	120	129	0.05%
ICD10	11,529	5	5	12	0.10%

4 Mapping of Guidelines to Find Relevant Concepts

In this method, we used clinical guidelines a source of information about domain related concepts, in order to identify a disease-centric subdomain of an ontology. Medical guidelines describe recommendations and conclusions regarding proper treatment based on scientific evidence. They aim to reduce the growing gap between knowledge and the actual practice. In our research, we used breast cancer guideline developed by the joint initiative of the Dutch Institute for Health care Improvement (CBO) [3].

From formalised models of the guideline [8] we extracted the names of all treatment plans, as well as all parameters describing patient data and their possible values in case of enumerated types. The parameters either specify plan preconditions and intentions or data that can be requested from external sources during guidelines execution. We mapped extracted concepts manually and had it verified by medical expert. Then we used the obtained mapping as a gold standard to compare with the results which could be produced by automatic mapping tool, the MetaMap [2].

Practical experiences with manual mapping. The main challenges of mapping concepts extracted from the guidelines to SNOMED CT concepts were searching among the hundreds of thousands of SNOMED CT concepts for the equivalences. Mapping required understanding the meaning of concepts used in the guidelines and knowing the exact context where they were used. After the initial mappings were identified, we consulted with our clinical expert and made adjustments where necessary. Below we illustrate some of the difficulties which we encountered.

In many cases guidelines and SNOMED CT use different terminology to express the same information. 'Axillary-node-dissection-proper' used in the guidelines and 'Excision of axillary lymph node' defined in SNOMED CT are an example of such case. Finding corresponding concepts was done using key words or using synonyms found in medical dictionaries. In cases where both approaches failed, we checked the context in the guidelines or looked for an explanation of concepts in other resources. This applied in the case of abbreviations as well as full phrases.

On the other hand finding an exact lexical match can be sometimes misleading. Such a situation was encountered when the plan 'Mastectomy' was analyzed. In the guidelines it covers the plan 'Mastectomy-proper' and also other procedures such as 'Radiotherapy-chest-wall' and 'Breast-reconstruction'. Hence the plan 'Mastectomy-proper' rather than 'Mastectomy' should be mapped to the SNOMED CT concept 'Mastectomy'. Therefore knowing the context was necessary.

Differences in granularity and abstraction level caused most of the missing matches. This issue appears mostly in the case of multiterms concepts, which are commonly used in the guidelines. Examples of such compound concepts are therapy + drug e.g. anthracycline-chemotherapy-manual, or therapy + drug + number of repetition e.g. six-courses-anthracycline-chemotherapy. Multiterms concepts are also used to define the intentions of therapies, for example 'elimination-distant-metastases'. Such specific concepts turned out to be very unlikely to be found in SNOMED CT ontology.

In a few cases even the large SNOMED CT ontology is not expanded enough yet. For example, SNOMED CT contains no concept corresponding to the parameter 'patient-prefers-bct', describing the patients preference of breast conserving treatment over mastectomy.

All these points above show that the method of obtaining relevant subdomains by mapping from guidelines is essentially a manual operation that cannot easily be automated. Results of manual mapping are significantly better, our early work in this domain ([12]) also corroborate this.

Results of the manual mapping. We found around 60 exact matches (matches with the same meaning but not necessarily the same name) out of all 150 parameters extracted from the guidelines. In the case of treatment procedures, we found around 40 exact matches out 190 procedures, and 40 matches, where SNOMED CT concepts have a close but more general meaning. The missing matches are caused by the reasons mentioned above.

Benchmark against MetaMap. In order to verify that manual mapping is indeed necessary we tested the applicability of MetaMap tool for the purpose of our research. MetaMap is a program developed at the National Library of Medicine to map biomedical text to the Metathesaurus [2]. It combines computational linguistic techniques with symbolic, natural language processing. MetaMap performs mapping in five main steps:

1. Parsing. The entire text is parsed, and divided into simple phrases using the SPECIALIST minimal commitment parser [9] which produces a shallow syntactic analysis of the text.
2. Variant Generation. In second step for each phrase are generated variants, using SPECIALIS lexicon and a database of synonyms, including all acronyms, synonyms, derivational and spelling variants of the given phrase.
3. Candidate Retrieval. Further the algorithm retrieves the set of all Metathesaurus strings, containing at least one of the variants.
4. Candidate Evaluation. Retrieved candidates from Metathesaurus are used to generate the mappings, which are evaluated using a linguistically principled evaluation function consisting of a weighted average of metrics measuring centrality, variation, coverage and cohesiveness. Then the list is ordered according to calculated scores.
5. Mapping Construction. Final mapping are constructed by combining candidates involved in disjoint parts of the phrases, and evaluated using the same scoring function. The mapping with the highest score is the best proposal of MetaMap.

We tested MetaMap on the same set of parameters and treatment plans extracted from the Breast Cancer guidelines. We compared the results with the results obtained by the manual mapping experiment. For each concept extracted from the guidelines, we checked whether its corresponding SNOMED CT concept, identified during manual mapping, is in the list of candidates proposed by MetaMap. In order to avoid ambiguity, and include equivalent mapping of different synonyms, we used for the comparison UMLS identifiers instead of concept names. It was possible due to the fact that SNOMED CT is included in the UMLS Metathesaurus. We tried different settings options to gain the deeper insight of MetaMap possibilities, including 'Term processing' and 'Ignore stop phrases'. In 'Term processing' mode input text is not divide into simple phrases but considered as a whole, which seems to be more adequate in the case of mapping concepts, which are most commonly multiword concepts.

The biggest overlap between the results produced by these two different mapping methods contains 30 out of 190 treatment plans and 16 out of 150 parameters. It was obtained using 'Term processing' mode. When for the comparison were used only the best candidates of MetaMap algorithm, then the numbers of exactly the same mappings decreased to 22 and 14 in case of plans and parameters respectively. Obtained results are summarized in table 3.

The major reason for differences in obtained mapping result are different strategies used for dealing with multiterms concepts. MetaMap proposes list of individually mapped Metathesaurus concepts, whereas we were aiming for finding a single corresponding concept with the closest meaning. For example MetaMap suggests for the concept 'Tumour negative excision margins' following mapping : 'Tumor excision NOS', 'Negative', 'Margin (Marginal)'. Manual

Table 3. Comparison of results of identified SNOMED CT concepts obtained using different mapping strategies

Mapping strategy	Identified parameters	Identified plans
MetaMap (all)	16 (10%)	30 (15%)
MetaMap (first)	14 (9%)	22 (12%)
Manual	60 (40%)	80 (41%)

browsing of the ontology and awareness of the application context let us identify the actual corresponding SNOMED CT concept 'Breast surgical margin not involved by tumour'. Automatic identification of such lexically unrelated concepts could be possible if for example SNOMED CT contained rich enough list of synonyms.

Results obtained using MetaMap, 15% of plans, and 10% of parameters correctly mapped to single corresponding concepts, clearly show that using mapping tools, which focus on lexical matching, is not sufficient in case of text composed using non-standard terminology, as that provided by SNOMED CT. It confirms our concern that manual mapping is necessary manual exercise in such case.

The set of identified SNOMED CT concepts, obtained by mapping guidelines concepts will be further expanded as described below.

Results of the expansion steps. In section 3, seed queries were used for the lexically querying for matching concepts. In the guideline-based method, this step is performed more semantically, namely by manually mapping the parameters and procedures of the guideline. In both cases, this first step is followed by subconcept-based expansion (transitively including all subsuming concepts, fig. 2) and UMLS expansion (using UMLS to include equivalent concepts, fig. 3).

Applying these two expansion steps to the results of the first manual mapping step resulted in an expansion from 140 to 2250 concepts. The two expansion steps have a much bigger impact after the manual mapping (from 140 to 2250) than they have after the first step in the seed query method (from 198 to 279). This difference can be explained by the fact that the first step in the seed query method returns mostly very specific SNOMED CT concepts that have very few subconcepts, while the manual mapping also yielded concepts higher in the SNOMED CT hierarchy.

However, also in the manual mapping case, the breast cancer-centric subdomain is again a very small fraction of the entire ontology, namely 0.73 % of the full ontology (308.677 concepts).

5 Evaluation of the Two Methods

Our two methods for identifying breast cancer-centric subdomains provided different results. The manual guideline-method found 2250 concepts, against 279 concepts found by the seed query method. Of these 279 concepts, 155 are

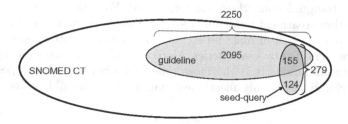

Fig. 4. Breast cancer subdomains identified using different approaches

also found by the guideline-method. The inclusion relations are summarised in figure 4.

Unsurprisingly, all 2250 concepts found by the guideline-method are indeed relevant concepts for the breast cancer-centric subdomain, in other words this method has a high precision. This is unsurprising because all concepts are either direct mappings from parameters or procedures in the recommendations of a national breast cancer guideline, or are subconcepts of these concepts.

More interestingly, manual inspection of the 279 seed query results shows that this method has a near perfect precision (i.e. all the concepts it finds are indeed key concepts for the breast cancer-centric subdomain). This confirms the main hypothesis put forward in section 2.

The figure also shows that besides its high precision (finding only key concepts), the seed query method has a rather low recall: it finds less than 10% of the concepts found by guideline-method. This is to be expected since the seed query method is tuned to find only key concepts (instead of finding all relevant concepts). However, inspection of the 2095 concepts that are only found by the guideline-method reveals that there are quite a few key concepts still contained in that set. Hence, even when counting only key concepts, the seed query method has no perfect recall. Examples of obvious concepts that we found missing are "Breast surgical margin involved by tumor", very detailed concepts such "Metastasis in internal mammary lymph nodes with microscopic disease detected by sentinel lymph node dissection but not clinically apparent" and quite a few others.

Finally, and contrary to our prediction, the seed query results are not a subset of the results from the guideline-method. In fact, well over 40% of all seed query results (124) are not found by the guideline-method. Inspecting this set yielded the following explanations for this falsification of our hypothesis:

Guidelines do not cover diagnostic concepts: The biggest part of concepts in this group describe breast neoplasm in general, e.g. 'Carcinoma in situ of female breast'. The guidelines are focused on recommendation for treatment of already diagnosed breast cancer, which is malignant. Benign neoplasm is not broadly discussed, since such concepts would be rather covered by diagnosis guidelines.

Only the guidelines recommendations were used: Some of those concepts are connected with breast cancer but are not included in the recommendations, the only part of the guidelines which was formalized. For example recommendation do not mention treatment procedures for male breast cancer, whereas concepts like 'Carcinoma in situ of male breast' or 'Carcinoma in situ of areola of male breast' were identified by seed queries. Moreover guidelines predominantly focus on ductal carcinoma as it is the most prevalent disease. Other types such as lobular carcinoma were only mentioned marginally - and did not occur in the formalized version.

The guidelines do not mention procedures that vary between hospitals: In The Netherlands, some hospitals employ special oncology nurses for home care of patients, others don't. The national guidelines do not discuss procedures for which there is an accepted high local variance between hospitals.

Between them, these reasons would remove a substantial part of the outlying 124 concepts, although we are currently not able to determine the exact number.

6   Related Work

Identification of a disease centric subdomain out of a large medical ontology to some extent resembles the problem of ontology modularization [4] which is applied in the context of combining existing ontologies by importing relevant modules. While not exactly the same[7], our notion of a subdomain can be compared to the ontology module as defined in [4] and hence we consider the papers presented below as related work.

Existing methods in the literature often rely on an a priori modularization of the vocabularies. These are typically based on some notion of semantic distance, or on the connectivity-graph of the ontology [6,13]. Such methods providing uncustomized modularization do not fulfill the requirement which we are aiming to meet, identifying subdomains specific for a particular use of a vocabulary. However, the methods which create partitioning of an ontology based on a given signature can be an alternative to the presented here seed query method. We will have a closer look to some of them. Generally, modularization techniques are divided into prescriptive and analytic. In prescriptive approach, the user explicitly states what is in or outside of the module, which requires the changes in the syntax and semantics of the language. In [4] one can find many arguments against it. The authors stress the fact that consequently whole infrastructure, OWL reasoners and parsers have to be changed as well. This approach leads to the tight, non-standard solutions, which severely restricts the reusability by other organization. Therefore we will focus on description of techniques based on analytic approach, where ontologies are defined using standard syntax and semantics of OWL, and ontology tools provide modularization services. In [4] they are evaluated according to the aspect of module correctness (any inference deduced from the module should be deduced from the original ontology), module

---

[7] In our case, we do not impose all formal properties that a module has as it is not necessary in our target application.

completeness (a module should contain all relevant information, so the user can not recognize that not whole ontology is imported) and module minimality (a module should be as small as possible).

One of analytic, ad hoc solution can be produced using PromptFactor algorithm [10]. It extracts a fragment of an ontology, based on a given signature. Created modules contain axioms that are mentioned in that signature and are further expanded with other concepts mentioned in those axioms until a fixed point is reached. The algorithm has been evaluated in [5], where authors prove that it is not always complete and creates modules larger than those, created by other algorithms that can guarantee completeness.

CEL and MEX are algorithms which work only with tractable fragments of OWL, the EL family of DL. This restriction is not problematic in case of SNOMED CT ontology, but NCI thesaurus and GALEN are beyond expressiveness of EL. The CEL reasoner [14] provides modularization technique based on connected reachability. Reachability can be expressed by a graph, where nodes are labeled with concepts from the ontology and edges are labeled with axioms. The modules contain the concepts themselves and the concepts and axioms from labels of connected nodes. They are guaranteed to be complete. MEX [7] can be applied only for acyclic EL ontologies, it generates minimal modules, smaller then other more generic algorithms.

Locality based algorithms are seen as most promising ones. Informally axiom is local if it does not change the meaning of concepts if included in the module. Changes of meaning are recognized differently according to the chosen locality type, e.g. axiom is top-local for a class if it does not define a new subclass for the concept. Basing on the application one can choose top or bottom locality, to be able to effectively generalize or refine set of identified axioms. Produced modules are proven to be correct and complete and the empirical analysis described in [5] attests the better approximation of minimal modules then other known algorithms.

7 Summary and Conclusions

Summary. Medical vocabularies are typically very large, containing up to hundreds of thousands of concepts. However, for any particular usage of such vocabularies only a small fraction of the concepts will be needed. In our example use-case, the breast cancer-centric subdomain of SNOMED CT is at most 1% of all concepts in the ontology. This gives urgency to the question of how to find such relevant subsets of concepts from potentially very large vocabularies.

In this paper, we have investigated two methods for identifying such relevant concepts. Our first method consisted of manually identifying a number of seed queries, and performing a lexical search for all concepts whose lexical labels contain any of the seed queries as a substring. All of the resulting concepts and their subconcepts are then considered as relevant for the subdomain characterised by the seed queries followed by the expansion phases.

Our second method consisted of extracting concepts that appeared as a parameter or procedure in the recommendations of the Dutch national guideline for the treatment of breast cancer. These concepts were mapped to SNOMED CT concepts. We compared the results of manual mapping with those obtained using MetaMap tool, which clearly showed that using automatic tool, which focus on lexical matching is not sufficient. This step was followed again by the two expansion phases.

These methods differ from other approaches for the identification of relevant subvocabularies that are based on any *a priori* modularization of the ontology, but instead select sets of concepts that are specific for a particular use of a vocabulary.

Conclusions. Our findings indicate that:
 - the breast cancer-centric subdomain is indeed only a fraction ($< 1\%$) of all concepts in SNOMED CT
 - the seed query method has a high precision, returning only key concepts
 - the seed query method has a low recall for returning relevant concepts
 - the guideline-method has a higher recall for relevant concepts while still having a high precision for relevant (but possibly non-key) concepts.
 - contrary to our prediction, not all key concepts are found by the guideline-method. Close inspection yielded a number of reasons why this is the case in our experiment:
   • the guideline covers only procedures for treatment, hence misses diagnostic concepts
   • we extracted our concepts only from the recommendations in the guideline, hence missing those concepts that only appear in the background information
   • the guideline does not mention procedures that vary between hospitals

Future Work. In future work, the validity of our conclusions should be tested by running these experiments on other subdomains (e.g. different diseases), and possibly using other methods to obtain a "gold standard" (our gold standard was obtained by manual extraction of all concepts from a national treatment guideline).

Similarly, it would be interesting to apply the guideline-method to other documents such as patient-records to see if that would yield a very different set of concepts.

More insight should be obtained in the correct choice for the seed concepts, since obviously the method is sensitive to this. The apparent fixed-point behaviour of this method deserves further investigation, for example on the degree of sensitivity to the initial set of query-concepts.

In addition we would like to take a closer look to various modularization algorithms. It would be very interesting to compare modules produced by different methods to learn more about their applicability for identifying disease specific concepts.

References

1. Aleksovski, Z., Vdovjak, R.: Overlap of selected ontologies in the context of the breast cancer domain. In: Proceedings of SIIM 2009 (2009)
2. Aronson, A.R.: Metamap: Mapping text to the umls metathesaurus. In: Proceedings AMIA Symposium (2001)
3. CBO. Guideline for the Treatment of Breast Carcinoma. van Zuiden. PMID: 12474555 (2002)
4. Clark, K., Parsia, B.: Modularity and owl (2008)
5. Grau, B.C., Horrocks, I., Kazakov, y., Satler, U.: Modular reuse of ontologies: Theory and practise. Journal of Artificial Intelligence Research (2008)
6. Cuenca Grau, B., Horrocks, I., Kazakov, Y., Sattler, U.: Just the right amount: extracting modules from ontologies. In: Proceedings of WWW, pp. 717–726 (2007)
7. Konev, B., Lutz, C., Walther, D., Wolter, F.: Cex and mex: Logical diff and semantic module extraction in a fragment of owl. In: Proceedings of the OWL: Experiences and Directions Workshop, OWLED 2008 (2008)
8. Marcos, M., Galan, J.C., Martinez, B., Polo, C., Seyfang, A., Miksch, S., Serban, R., ten Teije, A., van Harmelen, F., Rosenbrand, K., Wittenberg, J., van Croonenborg, J., Lucas, P., Hommersom, A.: Protocure ii deliverable d2.2bcd: Models of selected guideline in intermediate, asbru and kiv representations. Technical report (2005), www.protocure.org
9. McCray, A.T., Srinivasan, S., Browne, A.C.: Lexical methods for managing variation in biomedical terminologies. In: Proceedings of Symposium on Computer Applications in Medical Care, pp. 235–239 (1994)
10. Noy, N.F., Musen, M.A.: The prompt suite: interactive tools for ontology merging and mapping. Int. J. Hum.-Comput. Stud. 59(6), 983–1024 (2003)
11. Porter, M.F.: An algorithm for suffix stripping, pp. 313–316. Morgan Kaufmann Publishers Inc., San Francisco (1997)
12. Serban, R., ten Teije, A.: Exploiting thesauri knowledge in medical guideline formalization. Methods of Information in Medicine (to appear, 2009)
13. Stuckenschmidt, H., Klein, M.: Structure-based partitioning of large concept hierarchies. In: McIlraith, S.A., Plexousakis, D., van Harmelen, F. (eds.) ISWC 2004. LNCS, vol. 3298, pp. 289–303. Springer, Heidelberg (2004)
14. Suntisrivaraporn, B.: Module extraction and incremental classification: A pragmatic approach for el+ ontologies (2008)

Sharable Appropriateness Criteria in GLIF3 Using Standards and the Knowledge-Data Ontology Mapper

Mor Peleg

Department of Management Information Systems, University of Haifa, Israel, 31905
morpeleg@mis.hevra.haifa.ac.il

Abstract. Creating computer-interpretable guidelines (CIGs) requires much effort. This effort would be leveraged by sharing CIGs with more than one implementing institution. Sharing necessitates mapping the CIG's data items to institutional EMRs. Sharing can be enhanced by using standard formats and a Global-as-view approach to data integration, where a common data model is used to generate standard views of proprietary EMRs. In this paper we demonstrate how generic guideline expressions could be encoded in the GELLO standard using HL7-RIM-based views. We also explain how the Knowledge-Data Ontology Mapper (KDOM) can be used to simplify GELLO expressions. We are aiming to use this approach for computerizing radiology appropriateness criteria and linking them with EMR data from Stanford Hospital. We discuss our initial study to assess whether such computerization would be possible and beneficial.

Keywords: appropriateness criteria, clinical guidelines, GLIF, GEL, GELLO, EMR, ontology, knowledge sharing, KDOM.

1 Introduction

The American College of Radiology appropriateness criteria (AC) are evidence-based guidelines to assist referring physicians in making appropriate diagnostic imaging or treatment decisions. 147 AC are found in the National Guideline Clearinghouse (ngc.gov). Each set of AC addresses the diagnosis of one clinical problem (e.g., palpable breast mass) and recommends the radiological procedures that are suitable for different patient characteristics (variants). Each AC set contains 1-20 variants. For example, for diagnosing palpable breast mass, one variant is woman under 30 years of age who have palpable breast masses. For this population X-ray diagnostic mammography bilateral is recommended with a rating of 9 (which is the maximum rate) while two other radiological procedures are recommended with a lower rating of 8 (see Table 1) and Magnetic Resonance Imaging (MRI) of the breast is not indicated (has a rating of 2).

By employing AC, providers enhance quality of care by choosing the most appropriate procedures. However, as the AC are not in electronic form, it is difficult to ensure they are widely used in practice. Our aim is to encode AC and interpret them against patient data from electronic medical records (EMRs) in order to provide decision support on appropriate imaging or treatments.

D. Riaño et al. (Eds.): KR4HC 2009, LNAI 5943, pp. 64–75, 2010.

Table 1. Appropriateness criteria for palpable breast mass for the variant of women under 30

Radiological procedure	Rating
X-ray diagnostic mammography bilateral	9
X-Ray supplemental mammographic views	8
Ultrasound breast unilateral	8
MRI breast	2

Encoding and validating clinical knowledge, such as AC, is a labor-intensive task. Therefore, it would be useful to use a representation formalism that would support sharing the encoded knowledge among implementing institutions. If the knowledge contained in the AC (i.e., guideline knowledge) is expressed in a way that does not depend on the schema and terminology used in electronic medical records (EMRs) used in particular institutions, then the same encoding could be reused by different institutions. To enable execution of the generic guideline knowledge, patient data from the different EMR needs to be retrieved and abstracted to the same level of abstraction used in the guideline knowledge.

In this paper we present our approach to defining sharable guideline knowledge and to the simplification of the knowledge that is represented. We demonstrate our approach using GLIF3 [1] as the guideline modeling language used to represent the guideline knowledge. The paper is structured as follows. Section 2 provides related work. In Section 3 we discuss the methods used in this study: (a) the GLIF3 guideline modeling language, with its two possible expression languages: GEL and GELLO, (b) the HL7 Reference Information Model (RIM) [2] that can be used as a data model that bridges the knowledge of the guideline to the patient data schema of the EMR, and (c) Knowledge-Data Ontology Mapper, KDOM [3], which is an ontology and tool for mapping knowledge to data. In Section 4 we present an architecture for sharing guideline knowledge with different EMR systems. In Section 5 we show how KDOM can support simplification of the guideline knowledge representation. We provide the evaluation of our approach in Section 6. Section 7 provides a discussion.

2 Related Work

Representing medical knowledge such that the knowledge can be executed using existing EMR data and at the same time is sharable involves several challenges. First, the guideline modeling language needs to have a patient information model and an expression language that works with it. Guideline modeling languages that support such features include EON and GLIF [4] as well as the more recent SAGE [5] language. The Arden Syntax [6], although developed for modeling single decision rules, could also be used as a language for formulating guidelines [4]. In EON, GLIF, and SAGE, the patient information model is object-oriented and is based on the HL7 Reference Information Model, which is discussed in Section 3.2. In the Arden Syntax, the patient information model is very simple, and includes lists of time stamped data items.

The expression language is a central part of the guideline representation formalism. Such languages are used to formally represent clinical decision criteria that refer to patient data. For radiology AC, they the expression language is the most important feature of a guideline representation language. Expression languages should be expressive enough to represent different types of clinical expressions, including existence expressions (i.e., expressions that indicate existence of a condition, for example diabetes Mellitus), comparison expressions (e.g., systolic_blood_pressure > 120 mmHg), temporal expressions (e.g., latest cough lasting 4 weeks), and logical combinations of other expressions. In addition, expression languages need to be flexible in their use of different data structures, and be extensible such that more operators/functions could be added. Two expression languages have been standardized by HL7 include:

- The Arden Syntax [6], which is supported by commercial execution tools and used in clinical settings [4]. As mentioned above, the Arden Syntax works with a fixed data model which is not object-oriented. This inflexibility was one of the reasons for the development of the GELLO language [7].
- GELLO [8] is an object-oriented expression language that is an HL7 standard that can work with different data structures, is vendor independent and extensible. It is based on the Object Constraint Language (OCL) [9]. GELLO can easily be integrated with any HL7 RIM-based data model.

Fig. 1 shows examples of GEL and GELLO expressions.

```
(a) age < 30 years and gender = "woman" and palpable_breast_mass

(b) (PointInTime.now() self.player.oclAsType(livingSubject).
    birthTime) < ageThreshold and self.participation.act.
    oclAsType(observation).value ->select(code = '246188002' and
    codeSystemName = 'SNOMED-CT')->notEmpty()
    Where ageThreshold is defined as def: ageThreshold :
    PhysicalQuantity = '30 years'
```

Fig. 1. GEL/Arden Syntax and GELLO expressions. (a) The expression in Arden Syntax and in GEL (identical expression) for "woman under thirty with a palpable mass in her right breast"; (b) GELLO expression for "age < 30y".

Another challenge is the ability to share encoded knowledge by different institutions which use different EMRs. To support sharing, clinical criteria need to be specified using non-proprietary EMR codes. Instead, they should refer to standardized clinical terms taken from controlled vocabularies that are later mapped into concrete EMR fields. Much research has been done on facilitating this mapping of abstract guideline terms to concrete terms and EMR codes used in local implementations. Correndo and Terenziani [10] used the HL7 Common Terminology standard services to establish a link between a domain ontology and a database ontology in order to cope with heterogeneous term descriptions. This enabled them to use the GLARE modeling language in a way that is not committed to any specific ontology and database. The group of Shahar [11] developed the Medical Database Adaptor (MEIDA) tool that aids in linking knowledge-based medical decision-support systems to multiple clinical databases, using standard medical schemata and vocabularies. Their mapping tools use three heuristics: choice of a vocabulary to match the type of data item, choice of a key term, and choice of a measurement unit to narrow down the number of terms retrieved by the key term. An additional set of tools automatically maps standard term queries originating from the guideline to queries formulated using the local EMR's schema, terms and units.

Defining mappings between a guideline's patient data items and EMR fields needs to handle different types of discrepancies between the knowledge and data. The discrepancies include (1) mismatch in data model and terminology combinations [3], (2) use of abstractions by guideline authors, including (a) terms that need to be defined in terms of EMR fields (e.g., "breast mass" abstracts from raw data about particular locations of the mass on the right or left breast), (b) temporal abstractions, and (c) terminology abstractions (e.g., palpable breast mass is-a breast mass), and (3) differences in units of measure and time granularity [11]. Knowledge-Data Ontology Mapper (KDOM) [3] addresses the first two of these discrepancies. KDOM uses declarative query mapping supported by a mapping ontology defined in Protégé [12] and an SQL Generator that translates mapping instances into SQL queries used to retrieve the corresponding patient data. KDOM supports the definition of different kinds of abstractions using different types of mapping classes, including temporal mapping (e.g., first visit of a patient during 2004), hierarchical mapping (e.g., palpable breast mass and hard breast mass are kinds of breast masses), logical combination mapping (e.g., age<30 AND palpable breast mass), prior mapping, which allows nesting to support definition of complex mappings, and direct-one-to-one mappings that allow mapping an abstract term directly into an EMR or its view.

A problem that often arises in practical implementations of guidelines is lack of EMR data for clinical terms used in decision criteria. This problem has been observed and addressed in several studies [13, 14]. In some cases it is possible to add the necessary data items to the EMR [13]. Other times it is necessary to redefine the guideline's decision criteria so that they can suit the available local data [14]. In other cases, the abstractions used in the guideline are performed by the clinician, which either include 'holistic' assessments or abstraction rules that cannot be formalized with reasonable effort [13]. For example, rules for determining the aggressiveness of the tumor, or for deciding whether the patient is physically fit for chemotherapy.

3 Methods

In this work we chose to rely on standards. While there is no standard guideline modeling language, GLIF3 is a language that relies on standards, as explained below.

3.1 GLIF3

GLIF3 [1] is a guideline-modeling language that follows the task-network paradigm [15]. Guideline steps represent clinical actions, decisions, and action steps. These steps formally refer to patient data items, clinical concepts, and clinical knowledge. Branch and synchronization steps are used to enable parallel execution. To support sharing of guideline encoding by different institutions, GLIF3 uses standards, including controlled medical vocabularies and patient information model. Patient data items are specified by a *medical concept*, the code for which is taken from any controlled medical vocabulary (e.g., SNOMED-CT) and by a *data structure*, taken from a

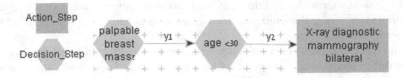

Fig. 2. A GLIF3 guideline corresponding to the appropriateness criteria of Table 1

standard reference information model (RIM), such as the Observation, Medication, and Procedure classes of the Health Level 7 (HL7) RIM [2]. Fig. 2 shows a simple[1] specification of the most recommended radiological procedure for women under 30 with palpable breast masses (Table 1).

GLIF3 has a formal language for expressing decision and eligibility criteria. In GLIF version 3.4, this expression language was the GEL [16] language based on the Arden Syntax [6], which is an HL7 standard. As an example, the GEL/Arden expression for "woman under thirty with a palpable mass in her right breast" is shown in Fig. 1(a). GLIF version 3.5 uses The GELLO expression language. We chose to use GELLO because it is an HL7 standard and can easily be integrated with any HL7 RIM-based data model. Unlike the Arden syntax, GELLO is suited for object-oriented patient information models, which simplifies writing criteria that relate to different properties of a concept. However, GELLO expressions could be quite complex. As an example, the GELLO expression for "age < 30y" is shown in Fig. 1 (b).

GLIF3 is supported by an execution engine [17], which is currently integrated with a GEL interpreter. Version 1 of GELLO is supported by an interpreter developed by the Australian company Medical Objects. That interpreter works with HL7 version 2 and 3 messages. An interpreter for version 2 of GELLO is being developed by Infer-Med. It will work with the HL7 Care Record model as the RIM-based data model and will align with OCL v2.1, which is about to be released by OMG.

3.2 HL7's Reference Information Model (RIM) and Virtual Medical Record

HL7's Reference Information Model (RIM) [2] is the primary interchange standard for clinical data both in the U.S. and internationally. The RIM, previously known as Unified Service Action Model (USAM) [18], provides a declarative way of specifying medical concepts and data items that are used in a guideline. The subset of the RIM used in GLIF3 contains the Act class and three of its subclasses: Observation, Medication, and Procedure, providing an information model for observations made about the patient, his prescribed medications, and medical procedures he underwent.

HL7 is currently developing a Virtual Medical Record (vMR). A vMR [19] provides a simplified RIM-based information model for patient data, enabling a guideline-based decision-support system to query a patient's state. HL7's vMR is being developed on the basis of the HL7 CareRecord refinement of the HL7 RIM.

---

[1] A more complex representation can use utility choices to represent all four possible radiological procedures using their ratings as utilities.

3.3 Knowledge-Data Ontology Mapper (KDOM)

We chose to use the KDOM mapping ontology and tool, discussed in Section 2 in order to map abstractions used in the guideline to EMR terms. KDOM is appropriate for this task and was evaluated to support mappings of a wide variety of guidelines; we previously used it to define all mappings necessary for linking the abstract terms defined in a GLIF3-encoded guideline for diabetic foot to patient data found in two different EMR schemas [3]. In addition, we found it sufficient for defining mappings from abstract terms contained in 15 GLIF3 encoded guidelines and one SAGE-encoded guideline into RIM views of these data items [3].

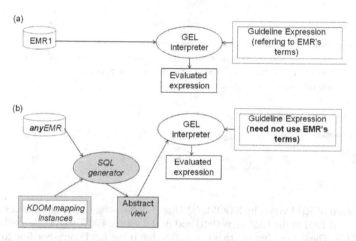

Fig. 3. Architectures for linking guideline knowledge to EMR data. (a) Direct linking of guide-line terms to EMR data. When the guideline refers to EMR terms the expression interpreter (GEL interpreter) can evaluate the guideline expression; (b) Mapping guideline knowledge to EMR data using abstract view of the EMR data using KDOM and its SQL generator. The SQL generator translates KDOM mapping instances into SQL queries. Running the queries in the EMR database management system produces abstract views of the EMR data. These views are stated using the terms that the guideline expression uses, which may be different than the EMR's terms. This enables writing the guideline expression using terms that abstract away from particular EMR implementations. The abstract view may (but does not have to) correspond to HL7's RIM model.

4 An Architecture for Sharing Guideline Knowledge with EMR Systems

Fig. 3 (a) shows an architecture for linking guideline knowledge directly to an EMR system. In this architecture, guideline expressions refer directly to EMR terms, making it possible for the expression language interpreter (e.g., GEL interpreter) to evaluate the expression against EMR data. Fig. 3 (b) shows how KDOM and its SQL generator can be used to enable separation between the guideline terms and the EMR terms. This enables encoding the guideline expression using terms that abstract away from particular EMR implementations.

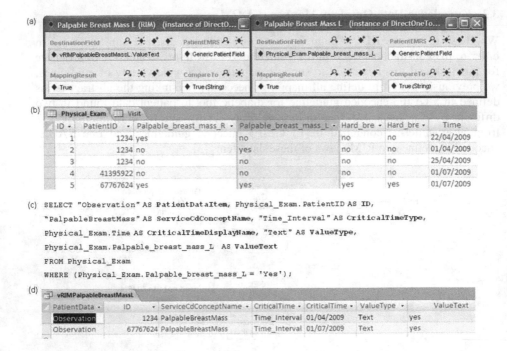

Fig. 4. Generation of RIM views by KDOM. (a) Direct-one-to-one mapping instances that indi-cate the tables and field in the RIM view (left) and in the source EMR (right), defined using the Protégé tool [12]. These mapping return a Boolean value if the field corresponding to palpable breast mass (in the RIM view or in the original EMR) holds the value "true" (represented as a String); (b) The EMR table "Physical Exam" and field "Palpable_breast_ mass_L (mass on the left breast) from which the RIM view is generated; (c) the SQL query used to generate the RIM views; (d) the RIM view produced by executing the SQL queries.

Using the architecture shown in Fig. 3 (b), KDOM can be used to implement the Global-as-View approach of data integration, where guideline expressions are mapped to EMR views in a common data model. To implement this approach, KDOM mapping instances of type Direct-one-to-one mappings (Fig. 4a) are used to access views of the EMR as an alternative of accessing a proprietary EMR. Fig. 4 shows how SQL queries (Fig. 4c) could be used to create RIM views (Fig. 4d) of a proprietary EMR (Fig. 4b). These queries are manually generated while the queries for defining abstractions over RIM views of the raw EMR data are automatically gen-erated by the SQL Generator based on mapping instances.

5   Using KDOM to Simplify Guideline Expressions

KDOM's mapping ontology can be used to define abstractions relating to simpler clinical concepts. By defining abstractions using KDOM, the guideline expressions that are defined in a guideline expression language such as GEL and GELLO can be simplified. For example, if the EMR includes two fields for specifying palpable breast mass –one for the right breast and one for the left– then KDOM can be used to generate a RIM view of palpable breast mass that combines the two separate fields using logical OR. In this way, the GELLO expression could be written in a generic way, as shown in Section 2.1, without the need to refer to the two fields separately. Moreover, KDOM can be used to integrate the EMR fields relating to palpable breast mass and to the patient's age into a single field: woman_under_30_with_palpable_ breast_mass, simplifying the guideline expression even further. The GELLO (or GEL) expressions refer to RIM views and not to actual EMR tables, allowing reuse of the mappings defined from guideline abstractions to RIM views. Changing of EMR data structure will not affect the original linkage of the guideline to the RIM view.

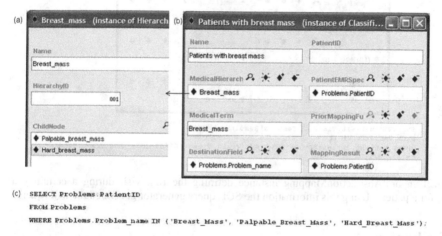

Fig. 5. Hierarchical mapping of breast mass using KDOM. (a) The hierarchy of breast masses defined in KDOM; (b) a classification hierarchy mapping instance that defines the abstract term corresponding to patients with breast mass. This mapping refers to the destination field Problems.Problem_name in an EMR. Through the MedicalHierarchy slot, this mapping instance defines the "Breast_mass" medical hierarchy (shown in part a) to be the hierarchy that specifies terms to be compared to the destination field. The MedicalTerm slot specifies that the search of the Breast-mass hierarchy should start with the root concept Breast_mass and visit its child nodes recursively. The result returned from the mapping would be the ID of patients with breast mass; (c) the SQL query generated by the SQL Generator using the knowledge defined in parts a-b.

The abstraction mappings supplied by KDOM include logical mappings (as in the example above), hierarchical mappings, and temporal mappings, which can be nested using the prior mapping class. All of these mapping classes could be used to simplify guideline expressions. For example, we can write a guideline expression that refers to breast masses and define in KDOM a hierarchical mapping that defines palpable breast mass and hard breast mass to be subclasses of breast mass. The SQL Generator would then be able to generate an SQL query that queries all these types of breast masses, as shown in Fig. 5.

Fig. 6 shows an example of a temporal abstraction mapping instance defined using KDOM and the SQL query generated by the SQL Generator, based on the mapping instance. While GEL includes temporal operators such as first and last, GEL's mapping interpreter does not support these operators. Therefore, KDOM can be used to define temporal abstractions, simplifying GEL expressions. For example, Fig. 5 shows a mapping instance defining the first visit during a certain year (2004). The results returned by this query could be bound to a variable called "first_visit_date" and a GEL criterion such as "first_visit_date > 2004-09-01" could thus be interpreted by the GEL interpreter. GELLO expression could also be simplified by referring to temporal abstractions defined in KDOM.

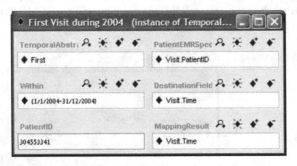

```
SELECT Min(Visit.Time) AS TemporalResult
FROM Patient, Visit
WHERE Patient.PatientID=304553341 AND Visit.Time>='01/01/2004' And Visit.Time
<='31/12/2004';
```

Fig. 6. TemporalAbstractionMapping instance defining the first visit during a certain year (2004) for a patient. Using this information the SQL query generator generated the SQL query.

6   Evaluation

We examined 44 of the 147 ACs (30%) to see whether they could be potentially represented in GELLO and whether data existed for these AC in the medical records used at Stanford Hospital. We first categorized the 44 AC according to the complexity of the criteria. We found that of the 44 AC, 9 were simple existence expressions (e.g., palpabale breast mass), 3 were expressions involving temporal operators (e.g., second trimester bleeding, recurrent UTI, chronic renal failure), 6 involved comparison (e.g., age < 65y), and 34 were complex expressions involving any of the other expression types. Note that some criteria that were complex also contained temporal or comparison criteria hence the number of criteria falling into the different categories do not add up to the total number of criteria in each category. The interesting result was that GELLO supported the expression of all of these criteria types.

However, the hard part was availability of data that the AC's noun phrases described. We examined the availability of data at Stanford Hospital. We differentiated between data that was found in structured form (e.g., pulsatile abdominal mass) and data that could be retrieved from radiology reports using natural language processing (NLP) techniques. For example, radiology reports of wrist exams include phrases that correspond directly to terms found in AC (e.g., distal radius fracture, scaphoid fracture, trauma) and phrases that would require inference (e.g., "fell off bike" or "motorcycle accident" that suggests trauma). As shown in Table 2, of the 44 AC, structured data was available for 5 criteria and data could be retrieved by NLP for 6 criteria, totaling in data availability for 25% of the AC. Data was not available for 33 of the 44 AC, for the following reasons. 14 noun phrases were too vague (see Table 2 for examples) and therefore naturally data for them did not exist in the medical records in structured form nor was there enough direction to suggest how NLP could be used to extract them from radiology reports. 4 noun phrases involved negative results that were not structured and could not be inferred with certainty from radiology reports. Finally, 20 AC contained phrases that were too detailed and often were not recorded in radiology reports or were recorded in such a varied way such that a fixed set of terms could not be defined to identify terms using NLP.

Table 2. Data availability and unavailability for ACs with different noun phrase types

Noun type	#AC	Example
Available data		
Structured	5 (11%)	Pulsative abdominal mass
Via NLP	6 (14%)	Wrist trauma
Total	11 (25%)	
Unavailable data		
Too detailed	20 (45%)	Suspect referred pain but wish to exclude hip
Vague	14 (32%)	Neurologic signs or symptoms present
Negative results	4 (9%)	Internal cervical os not visible by ultrasound
Total	33 (75%)	

7 Discussion

In this paper we demonstrated how the Global-as-View approach of data integration could be used to support guideline models that are encoded using abstract terms, which are mapped into a common global view of data arriving from various EMR formats. We further showed how KDOM could simplify authoring guideline expressions by defining clinical abstractions. The mapping of clinical abstractions to RIM views using KDOM could be reused when the same abstractions or data items are used in different criteria. When examining the ACs, we saw that many of the variants in a AC set refer to the same abstractions. Such reuse makes the work required to represent mapping instances more beneficial. While we demonstrated our approach using the GLIF3 language and its expression languages GEL and GELLO, this approach could potentially be used with other formalisms as well.

Our goal is to represent and share radiology appropriateness criteria. As we are interested in sharing knowledge, we prefer using a standard guideline expression language. Currently, the only standard guideline expression language that could be integrated with a standard object-oriented patient information model is GELLO, developed by HL7. Therefore, our intent is to use GELLO as the language for specifying appropriateness criteria and HL7's vMR as the common RIM model (data model) against which the GELLO criteria would be evaluated.

The characterization of criteria types as simple, complex, temporal, or comparison expressions makes it easier to find reusable patterns of representation in GELLO and in KDOM. Thus, after representing one type of criterion in GELLO or KDOM, similar criteria are represented similarly. The temporal operators recurrent and chronic could be expressed directly in GELLO, once their meaning is clarified (e.g., chronic meaning lasting over 3 weeks). Alternatively, KDOM could be extended to support these operators and simplify the encoding of GELLO criteria.

Lessons learned

We examined in detail 30% of the 147 ACs (44 AC) and saw that they could be potentially represented in GELLO; all the types of expressions, whether complex or simple, involving temporal operators or comparison operators could be represented. However, while technically it is possible to computerize radiology AC using the architecture and tools explained in this paper, what is needed is available EMR data. Unfortunately, data was available in structured form only for 11% of the criteria examined. Data for an additional 14% of the AC could potentially be retrieved from radiology reports using NLP techniques. Considering that a large effort is required to manually encode AC and map RIM views (corresponding to the vMR model) to EMR data and even more effort is required to integrate natural language processing, we are now assessing whether the effort required for AC automation, which could be done for just a small subset of the AC would be productive.

Acknowledgments. We thank Daniel Rubin for his help in analyzing the potential of computerizing the set of AC. We thank Robert Dunlop of InferMed and Peter Scott and Andrew McIntyre from Medical Objects for their help with the GELLO tool.

References

1. Boxwala, A.A., Peleg, M., Tu, S., Ogunyemi, O., Zeng, Q., Wang, D., et al.: GLIF3: a representation format for sharable computer-interpretable clinical practice guidelines. Journal of Biomedical Informatics 37(3), 147–161 (2004)
2. Health Level Seven. HL7 Reference Information Model (2006),
   http://www.hl7.org/Library/data-model/RIM/modelpage_mem.htm
3. Peleg, M., Keren, S., Denekamp, Y.: Mapping Computerized Clinical Guidelines to Electronic Medical Records: Knowledge-Data Ontological Mapper (KDOM). J. Biomed. Inform. 41(1), 180–201 (2008)
4. de-Clercq, P.A., Blom, J.A., Korsten, H.H., Hasman, A.: Approaches for creating computer-interpretable guidelines that facilitate decision support. Artif. Intell. Med. 31(1), 1–27 (2004)
5. Tu, S.W., Campbell, J.R., Glasgow, J., Nyman, M.A., McClure, R., McClay, J.P.C., Hrabak, K.M., Berg, D., Weida, T., Mansfield, J.G., Musen, M.A., Abarbanel, R.M.: The SAGE Guideline Model: achievements and overview. J. Am. Med. Inform. Assoc. 14(5), 589–598 (2007)
6. Hripcsak, G., Ludemann, P., Pryor, T.A., Wigertz, O.B., Clayton, P.D.: Rationale for the Arden Syntax. Comput. Biomed. Res. 27(4), 291–324 (1994)
7. Peleg, M., Boxwala, A.A., Tu, S., Zeng, Q., Ogunyemi, O., Wang, D., et al.: The InterMed Approach to Sharable Computer-interpretable Guidelines: A Review. J. Am. Med. Inform. Assoc. 11(1), 1–10 (2004)

8. Sordo, M., Ogunyemi, O., Boxwala, A.A., Greenes, R.A., Tu, S.: Software Specifications for GELLO: An Object-Oriented Query and Expression Language for Clinical Decision Support: Decision Systems Group Report DSG-TR-2003-02 (2004)
9. Object Management Group. Object Constraint Language
10. Correndo, G., Terenziani, P.: Towards a flexible integration of clinical guideline systems with medical ontologies and medical information systems. Stud. Health Technol. Inform. 101, 108–112 (2004)
11. German, E., Leibowitz, A., Shahar, Y.: An architecture for linking medical decision-support applications to clinical databases and its evaluation. J. Biomed. Inform. 42(2), 203–218 (2009)
12. Grosso, W.E., Eriksson, H., Fergerson, R., Gennari, J.H., Tu, S.W., Musen, M.A.: Knowledge Modeling at the Millennium (The Design and Evolution of Protege-2000). In: Gains, B.R., Kremer, R., Musen, M. (eds.) The 12th Banff Knowledge Acquisition for Knowledge-Based Systems Workshop, Banff, Canada, pp. 7-4-1–7-4-36 (1999)
13. Eccher, C., Seyfang, A., Ferro, A., Stankevich, S., Miksch, S.: Bridging an Asbru Protocol to an Existing Electronic Patient Record. In: Workshop on Knowledge Representation for Health-Care: Patient Data, Processes and Guidelines, in conjunction with AIME, Verona, Italy (2009)
14. Peleg, M., Wang, D., Fodor, A., Keren, S., Karnieli, E.: Lessons learned from adapting a generic narrative diabetic-foot guideline to an institutional decision-support system. Studies in Health Technology and Informatics, 243–252 (2008)
15. Peleg, M., Tu, S.W., Bury, J., Ciccarese, P., Fox, J., Greenes, R.A., et al.: Comparing Computer-Interpretable Guideline Models: A Case-Study Approach. J. Am. Med. Inform. Assoc. 10(1), 52–68 (2003)
16. Peleg, M., Ogunyemi, O., Tu, S., Boxwala, A.A., Zeng, Q., Greenes, R.A., et al.: Using Features of Arden Syntax with Object-Oriented Medical Data Models for Guideline Modeling. In: Proc. AMIA Symp., pp. 523–537 (2001)
17. Wang, D., Shortliffe, E.H.: GLEE – A Model-Driven Execution System for Computer-Based Implementation of Clinical Practice Guidelines. In: Proc. AMIA Symp., pp. 855–859 (2002)
18. Schadow, G., Russler, D., Mead, C., Case, J., McDonald, C.: The Unified Service Action Model: Indianapolis: Regenstrief Institute for Health Care (1999)
19. Johnson, P.D., Tu, S.W., Musen, M.A., Purves, I.: A Virtual Medical Record for Guideline-Based Decision Support. In: Proc. AMIA Symp., pp. 294–298 (2001)

Analysis of the GLARE and GPROVE Approaches to Clinical Guidelines

Alessio Bottrighi[1], Federico Chesani[2], Paola Mello[2], Marco Montali[2],
Stefania Montani[1], Sergio Storari[3], and Paolo Terenziani[1]

[1] DI, Univ. Piemonte Orientale "A. Avogadro", Alessandria, Italy
[2] DEIS, Univ. Bologna, Bologna, Italy
[3] ENDIF, Univ. Ferrara, Ferrara, Italy
{alessio,stefania,terenz}@mfn.unipmn.it,
{federico.chesani,marco.montali,paola.mello}@unibo.it,
sergio.storari@unife.it

Abstract. Clinical guidelines (GLs) play an important role in medical practice, and computerized support to GLs is now one of the most central areas of research in Artificial Intelligence in medicine. In recent years, many groups have developed different computer-assisted management systems of GL. Each approach has its own peculiarities and thus a comparison is necessary. Many possible aspects can be analyzed, but a first analysis has probably to consider the GL models, i.e. the representation formalisms provided. To this end, Peleg and al. [4] have analyzed and compared six different frameworks. In this paper, we analyse also GLARE and GPROVE on the basis of the same methodology. Moreover, we extend such analysis by considering the tools and the facilities that GLARE and GPROVE provide to support the use of GLs. The final goal of our analysis is to exploit the differences between these two systems and if they can be fruitfully integrated.

Keywords: clinical guideline, computer-assisted guideline manager, guideline model, decision support, verification.

1 Introduction

Clinical guidelines (GLs) represent the current understanding of the best clinical practice. In recent years the importance and the use of GL are increasing in order to improve the quality and to reduce the cost of health care. Many different systems and projects have therefore been developed in order to realise computer-assisted management of GL (see, e.g., the collections [1-3]), and computer–based GL management is now one of the most central areas of research in Artificial Intelligence (AI) in medicine and in medical decision making.

From the point of view of the GL model provided by such systems, i.e. of the GL representation language, a comparison among some existing systems is described in [4]. Such analysis concerns six approaches: Asbru [5], EON [6], GLIF [7], [8], GUIDE [9], PRODIGY [10], and PROforma [11].

D. Riaño et al. (Eds.): KR4HC 2009, LNAI 5943, pp. 76–87, 2010.
© Springer-Verlag Berlin Heidelberg 2010

Since 2003, new approaches have been developed (eg. GPROVE[12], HeCaSe [13], Helen[14], SpEM[15]) and some others (e.g. GLARE[16], GASTON [17], SAGE[18]) was not considered in [4]. In this work, we extend this comparison by analysing also GLARE [16] and GPROVE [12]. The methodology described in [4] concerns mostly a review of syntactic features. Since we consider important the analysis of the GL models, on the other hand we consider relevant that a computer–based GL system supports the user-physicians. Thus we address the tools and verification techniques provided by GLARE and GPROVE as well.

GLARE (Guideline Acquisition, Representation and Execution) [16] is a domain-independent prototypical system to acquire, represent and execute GL, which has been built starting from 1997 by University of Piemonte Orientale in cooperation with Azienda Ospedaliera San Giovanni Battista in Turin, one of the largest hospitals in Italy, and has been successfully tested in different domains, including bladder cancer, reflux esophagitis, and heart failure.

GPROVE (Guideline PRocess cOnformance VErification framework) [19] is a set of tools for the specification and run-time/a-posteriori compliance verification of the observed GL behaviour. It is composed by a graphical process definition language called GOSpeL (Guideline prOcess Specification Language), by an automatic mapping/translation module towards a formal language called *SCIFF* [20], and by an operational counterpart (a proof procedure) of the SCIFF formalism, that is used to verify the compliance of a given execution with respect to the defined GL process. GPROVE has been built by University of Bologna and University of Ferrara, and, in the SPRING PRRITT project sponsored by Emilia Romagna region, it has been successfully tested with the Cancer Screening Guideline adopted by the sanitary organization of the Emilia Romagna region.

Main goal of our work is to study the feasibility of integrating GLARE and GPROVE in a new common framework. First step of such study is to analyse their GL models in order to define their similarities and their differences. Moreover, analysing tools and facilities they offer, we can understand what the new common framework can provided to the users; in particular through this analyses, we can understand if the tools provided by the two approaches are or not complementary and thus the potentiality of the new framework.

The paper is organized as follows. In Section 2 we present the comparison between the GLARE and GRPOVE GL models using the methodology presented in [4]; in Section 3 we extend the analysis taking into account also the tools that GLARE and GPROVE provide to support use and verification of GLs. Finally in Section 4 we address conclusions and future works.

2 Analysis of GLARE and GOSPEL GL Models

Peleg et al. [4] identified eight dimensions to compare different GL approaches. These dimensions regard two broad categories: structuring guidelines as plans of decisions and actions, and linking a guideline to patient data and medical concepts.

We compare GLARE and GPROVE/GOSpeL[1] using these eight dimensions, following the same methodology described in [4]. In order to ground the analysis on a concrete setting, we started our comparison by acquiring the GL for managing chronic cough developed by American College of Chest Physicians [21], that is a revised and updated version of the GL [22] used in [4]. In Figure 1 and Figure 2 we show how parts such GL have been modelled by using GLARE and GOSpeL.

We only provide a short description of the eight dimensions used in [4]; the interested reader is referred to [4] for a detailed discussion about them.

Dimension 1. Organization of guideline plan components: this dimension deals with how the system supports the decomposition of GLs into networks of component tasks (i.e. plans) and how it allows to express various arrangements of these components and their interrelationships.

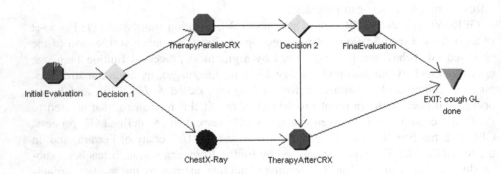

Fig. 1. Part of the chronic cough treatment guideline acquired in GLARE

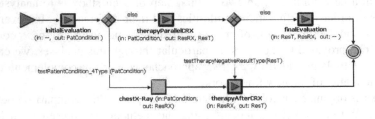

Fig. 2. Part of the chronic cough treatment guideline acquired in GPROVE

GLARE uses the single, generic construct "has-part" to define plans. Plans can be defined in terms of components (see dimension 3 for components types), which could be themselves plans. GLARE supports sequential, parallel, cyclical, and iterative plans. Parallelism is supported through the concurrency relation. Plan iteration can be specified by providing temporal constrains (e.g. maximum and minimum duration, frequency, periodicity, number of repetitions) and/or exit conditions. Although GLARE allows to define only one entry point for a GL, it is possible to specify multi

---

[1] GOSpeL is the representation language of the GPROVE framework.

entry point referring to patient states in decision criteria or preconditions that affect the guideline control flow.

GOSpeL has a single, generic construct to define a plan and supports nesting. GOSpeL supports sequential, parallel, cyclical, and iterative plans. To support parallelism GOSpeL provides two specific constructs: parallel fork and parallel join, which allow to define respectively the branching point of the multiple paths and synchronization point of the multiple parallel paths. Then GOSpeL provides two ways to define plan iteration trough two constructs: the *for* construct given the number of repetition and the *while* construct given a logical guard that works as an exit condition. Although GOSpeL allows to define only one entry point for a GL, it is possible to specify multi entry point using decisions having criteria referring to patient states.

Dimension 2. Specification of goals/intentions: this dimension deals with how the goals/intentions of actions can be defined.

GLARE specifies goals as text string; the goals are presented to the user-physician during the GL execution.

GOSpeL represents goals formally via SCIFF language [20] and uses these expressions to check compliance of a GL execution.

Dimension 3. Model of guideline actions: this dimension deals with the types and the characteristics of GL actions (i.e. the modeling primitives) used to represent the tasks described in GLs.

GLARE distinguishes between *composite* (see Dimension 1 for details) and *atomic* actions. GLARE provides four different types of atomic actions: *work actions*, *query actions*, *decisions* and *conclusions*. Actions are described in terms of their attributes following in ontology described in a set of dedicated databases. In **GLARE** the medical knowledge is stored in a set of databases; the system interacts with them providing the user-physician with information for the GL description.

In particular to specify action attributes, the user-physician can interact with the Pharmacological DB, that stores a structured list of drugs and their costs, with the Resources DB, that contains the resources available in a given hospital, with the ICD DB, that contains a coding system of diseases provided by the Azienda Ospedaliera, and with the Clinical DB, that provides a "standard" terminology to be used when building a GL. For what concerns temporal constrains, GLARE allows to specify qualitative and quantitative constraints, as well as duration of actions, delay between actions, repeated/iterative events; all types of constraints may be imprecise and/or partially defined (see [23] for details). For what concerns the exchange of information (i.e. see System Actions in [4] for details), GLARE models the patient data query, and during the execution the user-physician can specify the execution failure of one action.

GOSpeL describes a GL using *blocks*. The blocks are grouped into three families: *activities* blocks which represent guideline activities at the desired abstraction level; *gateways* blocks used to manage the convergence and the divergence of control flow; *start* and *end* blocks used to represent start and end points of (sub)processes. Activities can be *complex* (see Dimension 1 for details) or *atomic*. Atomic activities model a single atomic working step within the GL (i.e. a situation where a guideline participant should perform something). GOSpeL adopts an ontology-based approach to represent domain-related knowledge and its ontology is defined by two taxonomies:

one called *Activities,* which models activities at the desired abstraction level and divides them in administrative and clinical activities; a second taxonomy called *Entities*, which is used to describe domain's entities characterizing the guideline. Example of *Entities* mapped as *notLivingActor* are the healthcare structures, e.g., the general medicine, radiology and gastroenterology departments, and the different types of analysis results involved in the guideline. Each atomic activity block is semantically specified by mapping it onto an ontological activity and a set of participants. Temporal relations between blocks are expressed using *relations*, which represent causal binary connections between blocks and show how the flow navigates through blocks, imposing a partial ordering among them. Moreover in GOSpeL the temporal constrains between blocks can be defined as CLP constraints [24], allowing to express qualitative and quantitative constraints, as well as duration of actions, delay between actions, repeated/periodic events; all types of constraints may be imprecise and/or partially defined.

Dimension 4. Decision models: this dimension deals with the modeling methodologies for supporting decision making.

GLARE allows to represent two different kinds of decisions: *therapeutic* and *diagnostic* decisions. These decisions are not automatic, and thus the user-physician must always make her/his choice between alternative paths, since GLARE only shows information about whether a path is supported or not. In GLARE the criteria of a *diagnostic decision* are defined by a set of triples <diagnosis, parameter, score> (where a parameter is a triple <data, attribute, value>), plus a threshold to be compared with the different diagnoses' scores. Instead, in *therapeutic decisions* the decision is based on a pre-defined set of qualitative parameters: effectiveness, cost, side-effects, compliance, duration. Observe that in kinds of decision it is possible to express preferences for one or more alternative paths, that are presented in the guideline. Moreover, to support decision making, GLARE provides the decision theory facility (see Section 3 and [25] for more details). Observe that even if GLARE has not a switch construct, in which a decision is taken in a deterministic way, there are some types of automatic decision: the decision concerning the execution of an action with preconditions and of a cyclic action with exit conditions is taken deterministically by the system.

GOSpeL allows to define decisions using a *switch* construct called *exclusive choice*, in which the criteria are logical guard, where each guard is associated with an outgoing path and paths, mutually exclusive, are used to represent at design time all the expected alternative decision derivations. Moreover, GOSpel provides another kind of decision, called *deferred choice*, which models the decision autonomously taken by a participant (e.g. execute or not the PAP-test); this decision is nondeterministic and is not associated to explicit conditions on the alternative paths. Note that in GOSpeL the user-physician can not specify preferences for alternative paths with respect to a nondeterministic decision.

Dimension 5. Expression/criterion languages used to specify decision criteria: this dimension deals with the languages used to represent decision criteria, including pre- and post-conditions of GL plan components, and criteria that control plan execution states.

GLARE allows to express presence criteria (criteria concerning giving explicit definitions of terms to be checked). As described above, GLARE uses a threshold policy for what concerns diagnostic decision. Pre-conditions can regard the patient data (in this case they are defined as diagnostic decision criteria) and/or the presence of resources. Exit conditions are based on patient data (defined as diagnostic decision criteria), and the user-physician define whether all or at least n of them should be satisfied (where n is defined during the acquisition phase) to evaluate whether the exit condition of a cyclic action is satisfied during the execution. In GLARE the user-physician can not define temporal criteria concerning the time stamp of patient data and context-dependent expressions. However during the execution of GL the user-physician can decide whether a patient datum is reliable or not. Moreover GLARE does not provide *if...then...else* and *switch* statements in its expression language. Observe that GLARE provides templates to support the criteria definition.

GOSpeL models presence criteria by defining a Boolean data item on the data of the patient. This item is treated like all other data items and can be used in the decisional criteria. GOSpeL does not directly support pre- and post-conditions. They can be modelled using *exclusive choice* to tests conditions and to control the flow accordingly. Moreover, in GOSpeL it is possible to define criteria concerning the time stamp of patient data and context-dependent expressions. *If...then...else* statements can be modeled with *exclusive choice* as well: one outgoing connection of the *exclusive choice* construct is always associated with the *else* label, chosen if none of the conditions attached to the other connections turns out to be *true*.

Dimension 6. Data interpretations/abstractions: this dimension deals with the presence and the characteristic of abstractions, which aid in conceptualizing guideline logic and interpreting data. [4] identifies four types of abstractions: temporal abstractions/temporal patterns (trends), definitions of abstract terms, terminology abstractions via classification hierarchies, and scenarios and patient-state steps (discussed above in Dimension 1).

GLARE does not support temporal abstractions to abstract conditions that persist over time, based on raw, time-stamped values; it is possible to define new abstract terms, but these definitions are not based on formal expressions regarding patient data and/or other concepts (e.g. the user-physician can define the new *isolated systolic hypertension* concept, but s/he can not define its semantics as constrained by systolic blood pressures of at least 140 mmHg, by diastolic blood pressures less than 90 mmHg, and by the situation in which patients are not taking anti-hypertensive drugs). In GLARE the medical knowledge stored in a set of databases (see Dimension 3) follows a taxonomy-based organization via classification hierarchies.

GOSpeL does not support temporal abstractions too. The medical knowledge is organized in an ontology (see Dimension 3) formed by two taxonomies built using the Protégé tool [26]. The users can define new abstract terms, which can also be based on formal expressions regarding patient data and/or other concepts (e.g. the user-physician can define the new *isolated systolic hypertension* concept, and can define its semantics as constrained by systolic blood pressures of at least 140 mmHg, by diastolic blood pressures less than 90 mmHg, and by the situation in which patients are not taking anti-hypertensive drugs).

Dimension 7. Representation of a medical concept model and its use: this dimension deals with how medical concept can be represented in a model and then used.

In **GLARE** the medical knowledge is stored in a set of databases (see Dimension 3); the system interacts with them providing the user-physician with a "standard" vocabulary.

GOSpeL adopts an ontology-based approach to represent clinical knowledge; the user-physician can interact with two taxonomies: *Activities* and *Entities* (see the description in Dimension 3).

Dimension 8. Patient information model: this dimension deals with how the patient information model is defined. This model concerns also the definition of terminologies and of the structure of patient data.

In **GLARE** the patient data model is defined in the Clinical DB (see Dimension 3), which provides a "standard" terminology, and stores the descriptions and the set of possible values of clinical data. The patient data are stored in a specific Patient DB.

In **GOSpeL** a taxonomy is used (see Dimension 3) to describe domain's entities, namely actors, objects and terms. Every element is defined and managed through the Protégé tool [26]. A specific module called *Event Mapping* can provide the connection with the patient database of the different healthcare organizations.

3   Analysis of Tools Provided by GLARE and by GPROVE

Obviously, the GL model is an important feature, but it is not the only one which can be investigated to characterize a GL computer-assisted manager. We believe it is important to analyze also the GL computer-assisted managers from the point of view of the tools provided, to support the user-physician at differing stages within the GL life-cycle. In this section, we show the results of our analysis concerning tools and facilities provided by GPROVE and by GLARE.

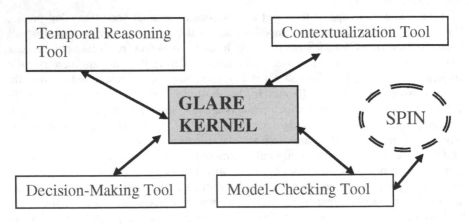

Fig. 3. The GLARE architecture

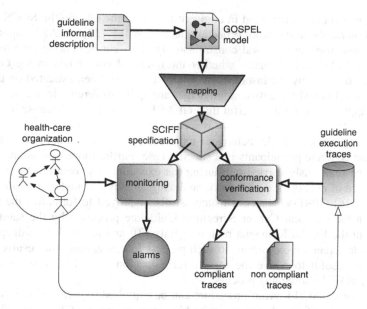

Fig. 4. The GPROVE framework

A first important difference between the GLARE and the GPROVE approaches regards their goals. The GPROVE (its architecture is shown in Figure 4), framework focuses on the compliance verification of the process executions with respect to the specified models, whereas GLARE (its architecture is shown in Figure 3) mainly focuses on GLs acquisition and execution. Of course, such a different focus is reflected in the different tools that they embed.

Both approaches provide a graphical editor that allows user-physicians to acquire and model GL easily. GPROVE proposes the GOSpeL Editor. In GOSpeL, a GL is composed by two parts: a flow chart, which models the process evolution trough a graphical language, and an ontology, which describes at a fixed level of abstraction the application domain and gives a semantics to the diagram. The ontology can be developed by using an ontology editor such as Protégé [26]. In GLARE, the graphical editor allows to aquire a GL by mean of primitives for drawing the control information within the GL and ad hoc windows to acquire the internal properties of the objects. Furthermore, during GL acquisition GLARE provides a set of facilities to check consistency and terminological correctness. For what concerns temporal constrains, GLARE provides a high-level language to easily represent the temporal action aspects. It then supports the possibility of checking, during acquisition, the temporal consistency of the GL, by exploiting a temporal reasoning tool, which operates in polynomial time on the number of GL actions [23]. In this phase, GLARE also provides a contextualisation tool to manage the gap between the generality of GL themselves (as defined, e.g., by specialists' committees) and the peculiarities of the specific application contexts. This tool allows to adapt GLs on the basis of the resources available in a given context (see [29] for details).

For what concerns compliance, the GPROVE framework is able to check the compliance of a given partial or complete history of a specific execution (i.e., the set of already happened events recorded in an event log) with the GL via the SOCS-SI tool [27], based on an abductive proof procedure named SCIFF [20]. GLARE supports the check of compliance on temporal constrains only. This check is done by a temporal reasoning tool [23] that evaluates whether the temporal constraints in the GL have been respected or not by the instances of actions that have been executed on the specific patient. Obviously the two approaches are quite different; in particular, the GPROVE approach is more powerful than GLARE in this setting, because it does not only check if temporal constraints are respected, but is able to reason about actions and their data, checking if the behaviour expected by the GL model is actually followed by the concrete participants in a specific case. Furthermore, compliance verification can be seamlessly carried out during the execution, by dynamically acquiring the occurrence of events, or a-posteriori, analyzing already completed executions. For example, in [16] GPROVE has been successfully employed to formalize the process described in the Cervical Cancer Screening Guideline proposed by the sanitary organization of the Emilia Romagna region of Italy [2] and to check the adherence of 1950 concrete screening executions to such process formalization. The results of such analysis were useful to revise the former formalization and identify some relevant characteristics of the screening process.

A variant of the SCIFF proof procedure can be exploited to perform static verification, aiming at identifying design errors and inconsistencies in a GL model. The same verification can be also done in GLARE, which is loosely coupled with the model checker SPIN [28].

Since actually GPROVE framework does not have an execution engine, it does not provide tools and facilities for managing the execution. On other hand, GLARE, which has an execution engine, considers decision making as a crucial issue and provides different facilities to support it. First of all it incorporates a set of facilities based on the temporal reasoning tool: the user-physician can perform queries to obtain temporal information concerning a GL specific execution and can use a simulation facility to see the temporal consequences of choosing among different alternative paths. Moreover via the decision making tool, GLARE provides a decision theory facility, which allows the user-physician to identify the optimal policy, and to calculate the expected utility along a path by exploiting classical powerful dynamic programming algorithms (see [25] for details), and allows also to calculate costs, time and resources required to complete paths in a GL (see [30] for details).

As illustrated above GLARE and GPROVE have different goals and their tools reflected these goals. In particular, the tools embed provided support in different part of lifecycle GL (as show in Table 1). In particular, only GLARE provides a set of facilities to user-physician during acquisition phase, and a Decision-making tool to support user-physician during the execution of a GL on a specific patient. Both GLARE and GPROVE support property verification: GLARE via a model-checking tool based on SPIN and GPROVE via SCIFF proof procedure. Instead only GPROVE provides a set of facilities, which regards the compliance verification of the process executions with respect to the specified models. In particular GPROVE can check the compliance of a given partial or complete history of a specific execution with the GL via the SOCS-SI tool, and provides a monitoring tool, which work on partial GL history. Table 1 proposes a synthesis of the comparison between GLARE and GPROVE facilities and tools.

Table 1. Synthesis of the comparison between GLARE and GPROVE tools

	GLARE	*GPROVE*
Acquisition	Graphical editor, checking consistency and terminological correctness, checking the temporal consistency, contextualization tool	GOspeL editor
Execution	Decision-Making Tool	---
Monitoring	---	SCIFF proof procedure on a partial execution
Conformance verification	---	SCIFF proof procedure on a complete or partial execution
Property verification	Model checking tool (via SPIN)	SCIFF proof procedure

4 Conclusions and Future Works

In this work we have compared the GLARE and GPROVE approaches using the methodology proposed in [4], and moreover by taking into account also the tools they provide.

In [4] the authors show that, for what concerns GL model, the six approaches analysed considering eight dimensions have both areas of considerable similarity and areas where different solutions have been adopted to face the same problem.

Thanks to the analysis presented in Section 2, we observe that GLARE and GPROVE propose a solution for almost all the eight dimensions and share with the six approaches analysed in [4] the similar basic features identified as requirements to define a standard for a GL modelling and management system. GLARE and GPROVE GL models have many similarities, especially for what regards the control-flow dimension, even if they slightly differ in the kind of GL knowledge they can represent: GLARE GL model provides a more easy and intuitive way to define the procedural knowledge whereas GPROVE GL models is more oriented to express declarative knowledge and constraints.

On the other hand, our study shows that the provided tools and facilities are quite complementary and focused on different goals (see discussion in section 3). GLARE provides a more rich set of tools than GROVE, mainly because the GLARE project has been started more than ten years ago, while GPROVE has been proposed only recently. In particular, GLARE provides features to support GL acquisition and to support user-physicians in the decision making process during the execution of GL on a specific patient, while GPROVE provides very interesting features for what regards run-time/a-posteriori compliance verification, thanks to the possibility of automatically mapping GOSpeL models to the SCIFF formal framework.

The similarities between their GL models pointed out in this work show the feasibility of integrating the two systems. The idea of common framework is also supported by result of study regarding the tools provided. As a matter of fact, their tools are complementary, and thus such new common framework will encompass all the GLARE facilities and also a powerful compliance verification module. We therefore consider this work as a first step towards such integration. The second step has been presented at AIME' 09 [31], where we have defined a framework based on the integration between GLARE and GPROVE in order to use SCIFF formal framework to evaluate the GL conformance. The next step will be to investigate how the mapping from GOSpeL to SCIFF could be extended and adapted in the context of GLARE.

Acknowledgments. We would like to thank Manuela Rossi for her contribution to the comparison between GLARE and GPROVE. This work has been partially supported by the FIRB Project TOCAI.IT.

References

[1] Gordon, C., Christensen, J.P. (eds.): Health Telematics for Clinical Guidelines and Protocols. IOS Press, Amsterdam (1995)

[2] Fridsma, D.B. (Guest ed.): Special Issue on Workflow Management and Clinical Guidelines. Journal of the American Medical Informatics Association 22(1), 1–80 (2001)

[3] ten Teije, A., Miksch, S., Lucas, P. (eds.): Computer-based medical guidelines and protocols: a primer and current trends. IOS Press, Amsterdam (2008)

[4] Peleg, M., Tu, S., Bury, J., Ciccarese, P., Fox, J., Greenes, R.A., Hall, R., Johnson, P.D., Jones, N., Kumar, A., Miksch, S., Quaglini, S., Seyfang, A., Shortliffe, E.H., Stefanelli, M.: Comparing computer-interpretable guideline models: a case-study approach. Journal of the American Medical Informatics Association 10(1), 52–68 (2003)

[5] Miksch, S., Shahar, Y., Johnson, P.: Asbru: a task-specific, intention-based, and time-oriented language for representing skeletal plans. In: Proc. 7th Workshop on Knowledge Engeneering Methods and Languages, pp. 9–20 (1997)

[6] Musen, M.A., Tu, S.W., Das, A.K., Shahar, Y.: EON: a component-based approach to automation of protocol-directed therapy. Journal of the American Medical Informatics Association 3(6), 367–388 (1996)

[7] Ohno-Machado, L., Gennari, J.H., Murphy, S., Jain, N.L., Tu, S.W., Oliver, D.E., et al.: The guideline interchange format: a model for representing guidelines. JAMIA 5(4), 357–372 (1998)

[8] Peleg, M., Boxawala, A.A., et al.: GLIF3: The evolution of a guideline representation format. In: Proc. AMIA 2000, pp. 645–649 (2000)

[9] Quaglini, S., Stefanelli, M., Cavallini, A., Miceli, G., Fassino, C., Mossa, C.: Guideline-based careflow systems. Artificial Intelligence in Medicine 20(1), 5–22 (2000)

[10] Johnson, P.D., Tu, S.W., Booth, N., Sugden, B., Purves, I.N.: Using scenarios in chronic disease management guidelines for primary care. In: Proc. AMIA Annu. Fall Symp., pp. 389–393 (2000)

[11] Fox, J., Johns, N., Rahmanzadeh, A., Thomson, R.: Disseminating medical knowledge: the PROforma approach. Artificial Intelligence in Medicine 14, 157–181 (1998)

[12] Chesani, F., Lamma, E., Mello, P., Montali, M., Storari, S., Baldazzi, P., Manfredi, M.: Compliance checking of cancer-screening careflows: an approach based on computational logic. In: ten Teije, A., Miksch, S., Lucas, P. (eds.) Computer-based medical guidelines and protocols: a primer and current trends. IOS Press, Amsterdam (2008)

[13] Isern, D., Moreno, A.: Distributed guideline-based health care system. In: Proceedings of 4th International Conference on Intelligent Systems Design and Applications, ISDA 2004, pp. 145–150. IEEE Press, Budapest (2004)

[14] Skonetzki, S., Gausepohl, H.J., van der Haak, M., Knaebel, S., Linderkamp, O., Wetter, T.: HELEN, a Modular Framework for Representing and Implementing Clinical Practice Guidelines. Methods Inf. Med. 43, 413–426 (2004)

[15] Dube, K.: A Generic approach to supporting the management of computerised clinical guidelines and protocols, PhD thesis, Institute of Technology, Dublin, Ireland (2004)

[16] Terenziani, P., Montani, S., Bottrighi, A., Molino, G., Torchio, M.: Applying artificial intelligence to clinical guidelines: the GLARE approach. In: ten Teije, A., Miksch, S., Lucas, P. (eds.) Computer-based medical guidelines and protocols: a primer and current trends. IOS Press, Amsterdam (2008)

[17] De Clercq, P.A., Hasman, A., Blom, J.A., Korsten, H.H.M.: Design and implementation of a framework to support the development of clinical guidelines. International Journal of Medical Informatics 64, 285–318 (2001)

[18] Berg, D., Ram, P., Glasgow, J.: SAGEDesktop: an environment for testing clinical practice guidelines. In: Proceedings of 26th Annual Conference of the IEEE Engineering in Medicine and Biology Society (IEBMS 2004), San Francisco, USA, vol. 2, pp. 3217–3220. IEEE Press, Los Alamitos (2004)

[19] Chesani, F., De Matteis, P., Mello, P., Montali, M., Storari, S.: A framework for defining and verifying clinical guidelines: A case study on cancer screening. In: Esposito, F., Raś, Z.W., Malerba, D., Semeraro, G. (eds.) ISMIS 2006. LNCS (LNAI), vol. 4203, pp. 338–343. Springer, Heidelberg (2006)

[20] Alberti, M., Chesani, F., Gavanelli, M., Lamma, E., Mello, P., Torroni, P.: Verifiable agent interaction in abductive logic programming: the SCIFF framework. ACM Transactions on Computational Logic 9(4), 1–43 (2008)

[21] Irwin, R.S., Boulet, L.S., Cloutier, M.M., et al.: Managing cough as a defense mechanism and as a symptom: a consensus panel report of the American College of Chest Physicians. Chest 114(2), 133–181 (1998)

[22] Irwin, R.S., Baumann, M.H., Bolser, D.C., Boulet, L.P., Braman, S.S., Brightling, C.E., Brown, K.K., Canning, B.J., Chang, A.B., Dicpinigaitis, P.V., Eccles, R., Brendle Glomb, W., Goldstein, L.B., Graham, L.M., Hargreave, F.E., Kvale, P.A., Zelman Lewis, S., McCool, F.D., McCrory, D.C., Prakash, U.B.S., Pratter, M.R., Rosen, M.J., Schulman, E., Shannon, J.J., Hammond, C.S., Tarlo, S.M.: Diagnosis and management of cough executive summary: ACCP evidence-based clinical practice guidelines. Chest 129, 1–23 (2006)

[23] Anselma, L., Terenziani, P., Montani, S., Bottrighi, A.: Towards a comprehensive treatment of repetitions, periodicity and temporal constraints in clinical guidelines. Artificial Intelligence in Medicine 38, 171–195 (2006)

[24] Jaffar, J., Maher, M.J.: Constraint logic programming: a survey. JLP 19-20, 503–582 (1994)

[25] Montani, S., Terenziani, P.: Exploiting decision theory concepts within clinical guideline systems: towards a general approach. International Journal of Intelligent System 21, 585–599 (2006)

[26] Protégé ontology editor, http://protege.stanford.edu/

[27] The SCIFF abductive proof procedure,
http://lia.deis.unibo.it/Research/sciff/

[28] Holzmann, G.J.: The SPIN Model Checker. Primer and Reference Manual. Addison-Wesley, Reading (2003)

[29] Terenziani, P., Montani, S., Bottrighi, A., Torchio, M., Molino, G., Correndo, G.: A context-adaptable approach to clinical guidelines. In: Fieschi, M., et al. (eds.) Proc. MEDINFO 2004, pp. 169–173. IOS Press, Amsterdam (2004)

[30] Terenziani, P., Montani, S., Bottrighi, A., Torchio, M., Molino, G.: Supporting physicians in taking decisions in clinical guidelines: the GLARE "what-if" facility. In: JAMIA Symposium supplement, pp. 772–776 (2002)

[31] Bottrighi, A., Chesani, F., Mello, P., Molino, G., Montali, M., Montani, S., Storari, S., Terenziani, P., Torchio, M.: A Hybrid Approach to Clinical Guideline and to Basic Medical Knowledge Conformance. In: Combi, C., Shahar, Y., Abu-Hanna, A. (eds.) AIME 2009. LNCS (LNAI), vol. 5651, pp. 91–95. Springer, Berlin (2009)

Semantic Web-Based Modeling of Clinical Pathways Using the UML Activity Diagrams and OWL-S

Ali Daniyal and Syed Sibte Raza Abidi

NICHE Research Group, Faculty of Computer Science, Dalhousie University, Canada
{daniyal,sraza}@cs.dal.ca

Abstract. Clinical Pathways can be viewed as workflows, comprising an ordering of activities with associated execution constraints. Workflow models allow formal representation, analysis and execution of workflows in the Clinical Pathways. We present a semantic web-based approach where the domain knowledge and the workflow model are modeled separately as ontologies, while the Clinical Pathway and the associated workflows are modeled as the instantiations of these ontologies. Our workflow model is based on the UML Activity Diagrams and OWL-S service ontology, and the execution semantics are based on Place/Transition Petri Nets. We demonstrate our approach by capturing the workflow of the Prostate Cancer Care Pathway.

1 Introduction

Clinical Pathways (CP) aim to coordinate the care process for a specific condition at the institution level. In essence, CP describe the functional knowledge pertaining to an institution's clinical practices in terms of time-sensitive and outcome-driven processes—represented as a combination of plans, tasks, decisions, resources and care providers—that essentially resemble a *workflow*. Execution of CP for clinical decision support is a complex activity and demands (a) knowledge modeling—i.e. modeling the domain knowledge and the CP's functional knowledge that describes workflows involving multiple resources and actors; and (b) definition of execution semantics. Researchers have argued that CP knowledge modeling requires a formal Knowledge Representation and Reasoning (KRR) framework, and an execution model is required for describing the execution semantics [13].

The Semantic Web (SW) framework offers interesting methods to execute CP as it provides (a) semantically-rich knowledge modeling and representation formalism in terms of ontologies; (b) reusability of the knowledge models, (c) neat separation between domain and functional concepts, yet their easy integration to describe the CP knowledge; and (d) reasoning mechanisms to execute the CP knowledge represented in ontologies. To execute CP, we propose a synthesis of SW and workflow modeling techniques—SW based ontologies capture the domain specific aspects of a CP, whereas workflow modeling techniques such as UML activity diagrams allow the translation of the procedural aspects of CP into formal workflow models that characterize an ordering of clinical tasks and their associated executional constraints. Furthermore, for integrating CP execution with Clinical Information Systems (CIS) we

D. Riaño et al. (Eds.): KR4HC 2009, LNAI 5943, pp. 88–99, 2010.
© Springer-Verlag Berlin Heidelberg 2010

leverage web-services technologies as they inherently offer communication standards for interrelating heterogeneous applications.

In this paper we present our SW-based approach for the representation, analysis and execution of CP workflows. In our framework medical domain knowledge is modeled as *domain ontologies* and workflow knowledge as a *workflow model ontology*—CP workflows are modeled as instantiations of the domain and workflow model ontologies. We have developed an OWL-based CP workflow model based on UML Activity Diagrams and OWL-S process model (represented as a service ontology) as both these approaches have well defined Place/Transition Petri Nets (PTN) based execution semantics [2,3]. We argue that the translation of CP workflow descriptions to PTN allows us to define execution semantics for our workflow model thus enabling us to (a) execute the CP workflows; and (b) analyze the modeled CP workflows for various correctness issues such as deadlock, reachability, liveness, safeness and boundedness. In summary, we are developing a single framework for (i) ontology-based modeling of CP workflows employing ontology-encoded domain knowledge, (ii) analyzing and executing the workflows, and (iii) integrating the CP workflows with CIS operations. We demonstrate our CP workflow execution framework by modeling and executing workflows for Prostate Cancer Care Pathways where the domain knowledge is encoded in an ontology presented by Abidi et al. [5].

2 Related Work

A number of approaches exist to address the knowledge modeling and execution modeling needs for computerizing CP. These approaches can be classified as: (i) approaches focusing on the structural and functional modeling of the CP knowledge, (ii) approaches focusing on the execution models and analyses of the workflows, and (iii) approaches focusing on the integration of workflows with existing CIS.

Peleg et al. [8], Ye et al. [10] and Tu et al. [6] employed ontology models for capturing the domain knowledge. Peleg et al. [8] use the BioWf model for representing the structural domain knowledge by employing Protege-2000 KRR framework. Protege-2000 is also employed by Tu et al. [6] for domain knowledge modeling, while Ye et al. [10] employs OWL for describing the domain ontology. Dominguez et al. [9] developed a Life Assistance Protocol (LAP) model for capturing medical and workflow knowledge.

The workflow model employed by Pelege et al. [8] is based on the workflow model of the Workflow Management Coalition (WfMC) while Petri Nets (PN) are used as the execution model, through which they were able to perform different types of analyses and answer questions related to workflows. Ye et al. [10] employed OWL-S for modeling workflows, while the workflow management is achieved through OWL-S and SWRL-Rule-based modeling approach to temporal relationships. Dominguez et al. [9] employ Timed Parallel finite Automata for modeling workflows while the execution model is based a Multi-Agents Systems approach.

Tu et al [6] focus on the modeling of workflows in CP that allows integration of multiple data sources and the CIS operations. Anyanwu et al. [7] propose the METEOR system that employs WPDL of WfMC to describe the workflows and develops standards for interoperation of disparate sites. Tu et al. [6] discuss the SAGE system that is based on the previous work on guideline modeling including Proforma, GLIF, Asbru, EON, GEM, GLIF3, GUIDE and PRODIGY. A comparison of these systems in terms of their expressivity and features can be found in [13].

PN have been studied extensively to capture the execution logic of CP and biological processes. PN enable different types of analyses on the workflows and different PN tools facilitate study of the workflows. Peleg et al. [11] study properties and dynamics of biological systems and care pathways using different PN tools. [8,11,12] have based their execution models on PN. Du et al. [12] propose a framework of CP Adaptive Workflow Modeling based on Extended Workflow Nets, which is an extension of PN.

3 Our Solution Approach

In our work, we build on the above research with an emphasis on the integration of the CP workflow within existing CIS through a services oriented approach that employs SW services standard—i.e. OWL-S. Our approach is distinct from the related approaches in the way it provides a single integrated framework for (a) modeling domain and structural knowledge as sharable and reusable ontologies, (b) modeling workflows separately from the domain and structural knowledge but using concepts from it, (c) formally analysing clinical workflows for a number of properties and behaviours, and (d) integrating CIS operations within the workflow descriptions. Although each of the above mentioned aspect is covered in the solution frameworks proposed in the literature, there is no single framework enabling all of these tasks. To execute and analyze CP, our solution approach is to (a) use SW-based methods for modeling CP workflow; (b) use the UML Activity Diagrams and the OWL-S process model to describe CP workflows; (c) use the OWL-S grounding to integrate CP workflows with CIS; and (d) use Petri Nets execution model to both execute and analyze CP workflows.

The first step to formally represent a CP is to develop a workflow model that captures both the domain and functional aspects of a CP. We have developed a CP workflow model by combining two workflow and process modeling approaches—i.e. the UML Activity Diagrams and OWL-S service ontology—through the import and extension mechanism in SW (as shown in figure 1).

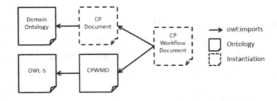

Fig. 1. Import hierarchy of documents for modeling workflows in CP

The UML Activity Diagrams provide an intuitive and expressive method to capture different workflow patterns for a variety of domains [1] with the provision of formal execution semantics [2]. We use UML Activity Diagrams to model CP because they are able to capture the complex, and at times nested, control-flow amongst multiple clinical tasks through standard constructs that represent complex workflow patterns, and the ordering constraints among the states. Since individual clinical tasks, modeled in the domain ontology, are now instantiated as independent UML Activity Diagrams constructs, we are able to re-use defined clinical tasks by combining or nesting them with other tasks to realize a functional CP workflow. Our approach for using UML Activity Diagrams for modeling CP tasks features re-usability of tasks through the separation of the institution-specific functional details from the domain knowledge—a generic clinical task can now be customized to meet local criterion by modulating the functional constraints.

OWL-S is a semantic web services standard that allows the linking of semantically compatible services based on WSDL descriptions. We leverage OWL-S to (i) enable integration of CIS operations as services within CP workflows, and (ii) capture pre-conditions, effects, inputs and outputs of the tasks in a CP. Our idea is to expose CIS operations as WSDL-based services which can then be integrated within workflow descriptions using OWL-S. It may be noted that, although OWL-S offers a standard process model to capture control-flow among sub-processes of a process, we use UML Activity Diagrams to model the CP process flow for two reasons: (i) The constructs of the OWL-S process model are less expressive for describing a complex control-flow as compared to the UML Activity Diagrams [4], and (ii) the UML Activity Diagrams offer a more intuitive approach to capture the ordering of tasks as direct relationships between tasks, as compared to the imposition of constraints practiced by OWL-S.

The use of UML Activity Diagrams and OWL-S model predicate PTN based execution semantics [2,3] which allows us to execute and analyze CP workflows.

To describe a CP workflow we import (a) RDFS/OWL based medical domain ontologies, and (b) CP Workflow Model Ontology (CPWMO) that entails the workflow knowledge. CPWMO is an OWL ontology of the UML Activity Diagrams which imports and extends the OWL-S service ontology. The execution model for CPWMO is based on the combined PTN semantics of the UML Activity Diagrams and OWL-S. We represent the workflow knowledge of a CP as an instantiation of the CP workflow model.

4   Ontology-Based Modeling of CP Workflows Using CPWMO

CPWMO is an OWL ontology for the UML Activity Diagrams importing and extending OWL-S service ontology. CPWMO ontology has classes for each of the UML Activity Diagram construct namely; ActivityDiagram, Initial, Final, ActionState, SubActivityState, Fork, Join, Decision, Merge, SendSignal and Receive-Signal. The ActionState and SubActivityState classes are described as subclasses of the proc:Process class, where proc refers to OWL-S Process Model ontology namespace. State is defined as the subclass of proc:Perform, the class of the instances of tasks and activities modeled as individuals of ActionState and SubActivityState respectively. A number of properties are used to capture ordering relationships among the UML Activity Diagrams constructs e.g. hasInitialState, hasFinalState, hasEdgeTo, hasCondition etc.

4.1 Modeling Actions and Activities in CPWMO

Atomic actions and complex activities are modelled in CPWMO as instances of the class State. Consider a domain ontology defining a task t in a CP P as an individual of a certain concept of domain. To define an atomic action that corresponds to a particular execution of the task t in a workflow associated with P, t is also declared as an instance of the class ActionState. A particular execution of the task t can then be modelled as an individual t' of the class State along with the assertion (t',proc:process,t). Modeling of tasks corresponding to complex activities that are themselves an ordering of a number of actions is achieved by employing the class SubActivityState and a particular execution of such an activity is modelled in a similar fashion. Note that the ordering constraints on the execution of actions in a workflow are modelled as statements about the individuals of the class State instead of the individuals of ActionState or SubActivityState. This approach allows reusing one task description in defining multiple workflows involving that same task.

4.2 Integrating CIS Operations in Workflow Descriptions

Our approach for integrating CIS operations with workflow descriptions is to leverage the grounding model provided in OWL-S for WSDL based web services. Our approach is to wrap CIS functions and operations as web services which are described in WSDL documents. Then using the OWL-S grounding model a mapping can be established between the semantics descriptions of services and the WSDL documents. Since both the ActionState and SubActivityState classes are subclasses of the proc:Process class in OWL-S, a grounding with concrete CIS operations can be described for each action in a workflow.

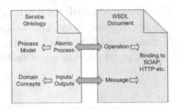

Fig. 2. Mapping between OWL-S grounding model and WSDL descriptions of CIS operations

Apart from address and protocol specific information required to invoke a concrete service, OWL-S grounding model also provides a mechanism of binding semantics descriptions of inputs and outputs of a process to a concrete service. This allows us to automatically invoke a CIS operation, while executing a workflow, with values to the inputs from within the workflows and then plugging in the values of outputs of the operation back into the workflows for further processing. Figure 2 illustrates this aspect of the OWL-S grounding model.

4.3 PTN-Based Execution Model for CPWMO

Our workflow model consists of the UML Activity Diagram constructs and the OWL-S service ontology. We employ the PTN based execution semantics for the analyses and execution of the workflows described using the workflow model. In the following we present the PTN based execution semantics for the UML Activity Diagrams and OWL-S process model that forms our execution model.

Place/Transition Petri Net. A PTN can be defined as a 4-tuple (P, T, I, O) where; P is a set of places, T is a set of transitions, $I:T\rightarrow 2P$ is an input map, and $O:T\rightarrow 2P$ is an output map.

It can be easily observed that a PTN can be represented as directed bi-partite graphs where the edges from places to transitions are determined by I and the edges from transitions to places are determined by O. Similarly any directed bi-partite graph can be viewed as a PTN. In a graphical representation of PTN places are represented by circles while the transitions are represented as boxes.

A marking $\pi:P\rightarrow\mathbb{N}\cup\{0\}$ of a PTN, with P as the set of places, is a mapping that can be seen as allocation of tokens to the places. The presence of tokens in the places results in the enabling of transitions which can then be fired to result in a change in the marking. A transition t of a PTN (P, T, I O) is enabled by a marking π iff $\forall p \in I(t), \pi(p) > 0$.

Let for a set Y and $X\subseteq Y$ define the membership function of X as $\mu_X: Y \rightarrow \{0,1\}$ as:

- $$\mu_X(y) = \begin{cases} 1, y \in X \\ 0, \text{otherwise} \end{cases}$$

Then we can define firing of an enabled transition t as a change in a marking π of a PTN (P, T, I, O) to another marking π' of the PTN as follows:

- $$\pi'(p) = \pi(p) + \mu_{O(t)}(p) - \mu_{I(t)}(p)$$

The transition firing mechanism of a PTN can be viewed as a simulation of the workflows defined as PTN. A token represents an active state whereas an enabled transition represents an action that is permissible due to certain states that are active. In this way firing of a transition can be interpreted as taking the permissible action which results in the change in active states of the system. In the following sections we discuss how PTN can be used to provide execution semantics for our CW model.

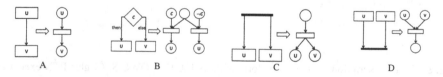

Fig. 3. (A) Translation of UML Activity Diagram constructs to PTN. U is a node of type I, A, M, J, S or R while V is a node of type F, A, D, F, S or R. (B) Translation of the decision construct to PTN. C is a condition while ¬ stands for logical negation. U and V are of type F, A, D, F, S or R. (C) Translation of the fork construct to PTN. U and V are of type F, A, D, F, S or R. (D) Translation of the join construct to PTN. U and V are of type I, A, M, J, S or R.

UML Activity Diagrams to PTN. We employ the PTN based execution semantics for the UML Activity Diagrams discussed by T. Ivana [2]. Figure 3 illustrates the translation of the UML Activity Diagrams constructs to PTN. For the sake of convenience we use a naming convention in which I, F, A, D, M, F, J, S and R stand for Initial node, Final node, Action State, Decision node, Merge node, Fork node, Join node, Send Signal node and Receive Signal node respectively. In figure 3 the circles in PTN represent places while the rectangles represent the transitions.

OWL-S service ontology to PTN. OWL-S captures three aspects of services, namely: Profile, Process Model and Grounding. The OWL-S process model provides with a number of constructs to define invocation order of a number of services to define a more complex process, called a composite process. We have employed OWL-S process model in our workflow model to capture preconditions, effects, inputs and outputs of tasks in CP (figure 4). We employ the PTN execution semantics of the OWL-S process model presented by S. Narayana et al. [3] (figure 5).

For modeling conditions in the UML Activity Diagrams we only consider SWRL conditions, that are conjunctions of RDF statements, in our workflow model. These are used to capture the conditions that decide alternative paths emerging from decision points in CP.

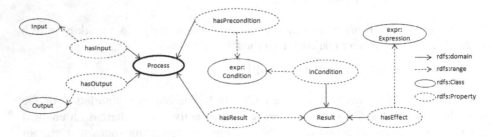

Fig. 4. Inputs, Outputs, Preconditions and Effects of services in OWL-S process model

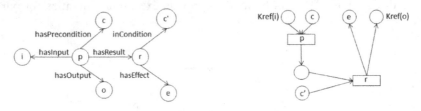

Fig. 5. (Left) Description of process p along with its IOPE in OWL-S, (Right) PTN execution semantics of the OWL-S given description

Translation of CPMWO workflows into PTN. Using the PTN based execution semantics of the UML Activity Diagrams and OWL-S service model, we first translate the workflows described using CPWMO into PTN and then perform safety, reachability and deadlock analyses on the workflow. Execution of a workflow is also based on the transition firing mechanism of the corresponding PTN.

5 Modeling Workflows in Prostate Cancer Clinical Pathways

Abidi et al. [5] presented the Prostate Cancer Clinical Pathways Ontology and described pathways for three regions—namely Halifax, Winnipeg and Calgary—as instantiations of the ontology. We refer to this ontology as PCONT in this paper. PCONT provides a subsumption hierarchy of 28 classes along with 34 properties to capture the clinical workflows and relevant medical knowledge. PCONT captures fine details about the classification of different types of tasks, decision criteria, treatments and actors. We use PCONT to demonstrate the working of our CP execution framework in terms of the modeling and execution of it.

As a first step, we translated the PCONT workflow information into the classes of CPWMO and captured the execution constraints in terms of the CPWMO ontology. Figure 6 shows a part of the prostate cancer clinical pathway encoded in PCONT, while figure 7 is the RDF graph of the CPWMO encoding of the corresponding workflow.

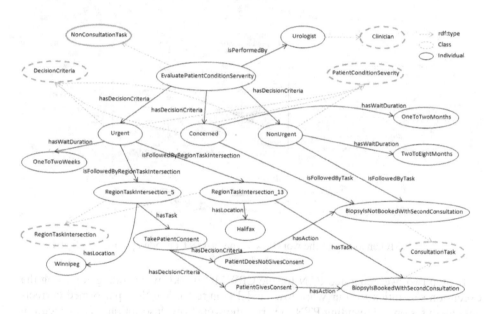

Fig. 6. RDF graph of a part of the PCONT encoded prostate cancer clinical pathway

The PCONT encoding of the pathway provides information related to the clinical tasks, information states, decisions, temporal information about the tasks and information about the pathway institutions. For example the encoding presented in figure 6 captures information related to tasks EvaluatePatientConditionSeverity and TakePatientConsent, and information states BiopsyIsBookedWithSecondConsultation and BiopsyIsNotBookedWithSecondConsultation.

To model the corresponding workflow information in CPWMO, instances corresponding to the above clinical tasks and information states are created as instances of the State class, and these instances are linked to the abstract tasks and states defined in PCONT using the proc:process. This approach allows us to separate descriptions of the tasks (defined in domain ontology) from the description of its invocation in a workflow, thus allowing re-use of the same task description in different workflows.

Decisions concerning the patient condition's severity, patient's location and patient's consent for biopsy are modeled by defining individuals of the class Decision, while merging of multiple paths into common paths is achieved by defining individuals of type Merge (figure 7). The conditions for decision making are captured as SWRL conditions. (?p hasConditionSeverity Urgent) is the condition for the for determining path leading from the EvaluatePatientConditionSeverity task, where ?p is a variable for patient that is bound to a patient in knowledge base at the time of invoking the pathway.

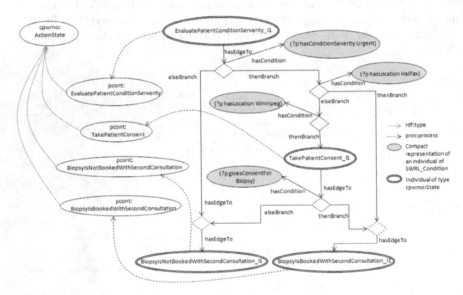

Fig. 7. CPWMO encoding of the workflow in the PCONT encoded pathway in figure 6

In step two, we generated PTN for the CPWMO workflow encodings by using the execution semantics for our workflow model (figure 8). We then performed correctness analyses on the resulting PTN to verify the workflow descriptions. The execution of the workflows is based on the transition firing mechanism of the PTN. For example after the execution of the EvaluatePatientConditionSeverity_I1, the transition leading to D_21 is enabled and therefore can be fired. Firing of this transition leads to the invocation of decision D_21. Now based on the patient's condition severity either one of the triples (?p hasConditionSeverity Urgent) or not(?p hasConditionSeverity Urgent) would hold, thus resulting in the invocation of D_22 or M_122 respectively.

We developed a Prostate Cancer Care Planning System that uses the execution model as its backend engine. The system updates the knowledge base according to the input provided by the user and queries the execution model for next state. Figure 9 shows snap-shots of our system for the two different executions of the part of workflow defined in figure 7. In the first case (left) the institution is in Halifax while the

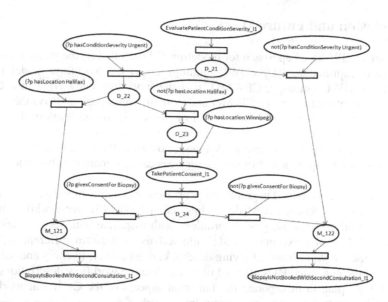

Fig. 8. PTN generated by translating the CPWMO-encoded workflow in figure 7

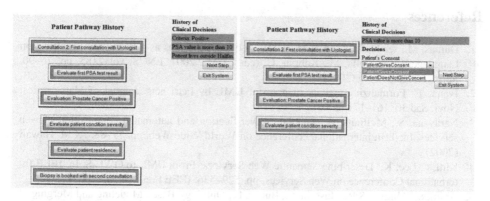

Fig. 9. Snap shots of the execution system demonstrating execution of the prostate cancer clinical pathways for different scenarios

patient lives outside of Halifax, while in the second case the institution is in Winnipeg and the execution is at a point where patient's consent for biopsy is required.

Using our approach we were able to formally represent the workflow in the prostate cancer care pathway using our CPWMO, leveraging domain knowledge encoded in PCONT. We were able to study correctness of the CP workflow by generating PTN from the workflow descriptions and then execute the CP by developing a system that leverages the execution model. In the next phase we are planning to integrate our system with CIS operations leveraging OWL-S grounding in WSDL.

6 Conclusion and Future Work

We presented a SW-based approach for modeling CP such that (i) the medical domain knowledge is captured as RDFS/OWL ontologies, (ii) the workflow model is described by an OWL ontology, CPWMO, (iii) the workflow knowledge in a CP is described as an instantiation of CPWMO and domain ontologies, and (iv) the execution model is based on the PTN. Our workflow modeling approach allowed us to (i) describe CP workflows by using concepts described in ontology-encoded domain ontologies, (ii) analyze and execute workflows, and (iii) at the same time integrate CP workflows with CIS operations exposed as services using semantic web services standard OWL-S.

Using the UML Activity Diagrams for workflow modeling provides an intuitive way of capturing workflow knowledge, while using OWL-S in our workflow model allowed us to address the integration problem with disparate data services and CIS operations. PTN based execution model allowed us to perform different types of analyses on the CP workflows. Studying deadlock, reachability, liveness and safeness properties of the generated PTN allowed us to validate the workflow descriptions.

In future we plan to incorporate the temporal aspects of the CP in our workflow model and develop a variance management framework.

References

1. Dumas, M., ter Hofstede, A.H.M.: UML Activity Diagrams as a Workflow Specification Language. In: Gogolla, M., Kobryn, C. (eds.) UML 2001. LNCS, vol. 2185, pp. 76–90. Springer, Heidelberg (2001)
2. Ivana, T.: Formalizing activity diagram of UML by Petri nets. Journal of Mathematics, Novi. Sad 30, 161–171 (2000)
3. Narayana, S., McIlraith, S.: Simulation, verification and automated composition of web services. In: 11th international conference on World Wide Web, pp. 77–88. ACM, Hawaii (2002)
4. Kim, I., Lee, K.: Describing Semantic Web Services: From UML to OWL-S. In: IEEE International Conference on Web Services, pp. 529–536. IEEE Press, Seoul (2007)
5. Abidi, S., Abidi, S.S.R., Hussain, S., Butler, L.: Ontology-Based Modeling and Merging of Institution-Specific Prostate Cancer Clinical Pathways. In: Knowledge Management for Healthcare Processes Workshop at 18th European Conference on Artificial Intelligence (ECAI 2008), Patras (2008)
6. Tu, S., et al.: The SAGE Guideline Model: Achievements and Overview. J. Am. Med. Inform. Assoc. 14, 589–598 (2007)
7. Anyanwu, K., Sheth, A., Cardoso, J., Miller, J., Kochut, K.: Healthcare Enterprise Process Development and Integration. Journal of Research and Practice in Information Technology, Special Issue in Health Knowledge Management, 83–98 (2003)
8. Peleg, M., Tu, S., Manindroo, A., Altman, R.: Modeling and analyzing biomedical processes using workflow/Petri Net models and tools. Studies in health technology and informatics 107, 74–78 (2004)
9. Dominguez, D., Fernandez, C., Meneu, T., Mocholi, J.B., Serafin, R.: Medical guidelines for the patient: introducing the life assistance protocols. Studies in health technology and informatics 139, 193–202 (2008)

10. Ye, Y., Jiang, Z., Yang, D., Du, G.: A semantics-based clinical pathway workflow and variance management framework. In: IEEE International Conference on Service Operations and Logistics, and Informatics, pp. 758–763. IEEE Press, Los Alamitos (2008)
11. Peleg, M., Rubin, D., Altman, R.B.: Using Petri Net tools to study properties and dynamics of biological systems. J. Am. Med. Inform. Assoc. 12, 181–199 (2005)
12. Du, G., Jiang, Z., Diao, X.: The integrated modeling and framework of clinical pathway adaptive workflow management system based on Extended Workflow-nets (EWF-nets). In: IEEE International Conference on Service Operations and Logistics, and Informatics, pp. 914–919. IEEE Press, Los Alamitos (2008)
13. Peleg, M., et al.: Comparing Computer-interpretable Guideline Models: A Case-study Approach. J. Am. Med. Inform. Assoc. 10, 52–68 (2003)

Extracting Qualitative Knowledge from Medical Guidelines for Clinical Decision-Support Systems

Maarten van der Heijden[1,2] and Peter J.F. Lucas[1]

[1] Institute for Computing and Information Sciences,
Radboud University Nijmegen, The Netherlands
[2] Department of Primary and Community Care,
Radboud University Nijmegen Medical Centre, The Netherlands

Abstract. Medical guidelines provide knowledge about processes that is not directly suitable for building clinical decision-support systems. We discuss a two-step approach where knowledge from a guideline on COPD is translated into temporal logic, and augmented with physiological background knowledge. This allows capturing the dynamics of the processes using qualitative knowledge, while maintaining the temporal nature of the processes. As a second step, this represented clinical knowledge is translated into a decision-theoretic framework. We thus present a representation that can act as a basis for the construction of a decision-support system concerning monitoring of COPD.

1 Introduction

Much knowledge of medical processes is available, but usually not in an easily accessible form. Representing this knowledge can be beneficial to make it more attainable. This is also recognised by the medical community which is becoming more and more evidence based. Medical guidelines are an attempt to systematically summarise what is known from clinical research. If such knowledge is represented formally, it is possible to automatically reason with it.

Recently, researchers have used temporal logic as a formalism to capture the temporal nature of knowledge available in clinical guidelines [1,2]. This allows reasoning with the represented knowledge for different purposes. Examples include checking formal properties such as reachability and whether particular treatment prescriptions are correct. We believe that such logical formalisations offer a good starting point for the construction of decision-support systems based on decision-theoretic principles. In this paper we investigate a formal knowledge representation for a decision-support system based on the guidelines on Chronic Obstructive Pulmonary Disease, COPD for short.

Patients with COPD suffer from a complicated progressive disease of the airways and lung, which has profound impact on the patient's well-being [3]. In the course of the disease there will be both gradual changes as well as more acute events, referred to as *exacerbations*. Early detection of a worsening of health status is beneficial to the patient and monitoring patients at home may accomplish early detection. Constructing an automated monitoring system requires

D. Riaño et al. (Eds.): KR4HC 2009, LNAI 5943, pp. 100–112, 2010.

decision-theoretic knowledge of COPD progression and treatment. We hypothe-sise that the qualitative information contained in medical guidelines can be used to fulfil these requirements at least partially. The problem we have to solve is how to represent the medical knowledge available from guidelines such that the representation facilitates decision making based on patient monitoring data.

This paper is structured as follows, Section 2 defines medical guidelines fol-lowed by a formalisation of the COPD guidelines in Section 3. Then in Section 4 we look at the decision-theoretic principles related to decision-support and finally conclude with a discussion.

2   Medical Guidelines

A medical guideline is a specification of detailed advice on treatment of a par-ticular disorder and is aimed at health care professionals. Medical guidelines in the Netherlands are issued by the Dutch Institute for Health Care Quality. The recommendations in the guidelines are based on medical literature, the quality of which is assessed and carefully rated by the working group that makes the guideline.

The document is structured such that for each topic the relevant literature is reviewed on quality (e.g. whether results come from a randomised trial) and on outcomes. This leads to a set of conclusions that are in accordance with the literature, each of them with a level of confidence. These levels follow directly from the number and quality of studies that support the conclusion. The recom-mendations are in turn a practical realisation of the conclusions supported by literature.

Guidelines are written in natural language – which has inherent ambiguity – and thus cannot be reasoned with automatically. We will create a more for-mal representation to overcome this problem. The guidelines do not deal with uncertainty in diagnosis or measurements, nor do they apply to every specific patient situation. The guidelines thus provide advice that should be followed in general, but the recommendations may be disregarded in specific cases if there are clinical reasons to do so according to the treating physician. Fig. 1 shows an example fragment from the COPD guideline.

3   Formalising the COPD Guidelines

Two guidelines have been issued on COPD treatment in the Netherlands. In this section we give a formalisation of these guidelines using temporal logic, which will result in a logic knowledge base. We will start with a short introduction to COPD, followed by some preliminaries on temporal logic, a formal domain description and the knowledge base construction.

3.1   Chronic Obstructive Pulmonary Disease

COPD is a high prevalence airway and lung disease, characterised by increas-ing airflow limitation, declining performance status and respiratory failure [3,4].

Conclusion (level 1):
Anticholinergics and β_2-agonists have comparable effect on bronchodilation by patients with an exacerbation. There is no evidence that their combination has a better effect than either one alone.
Recommendation:
During treatment of exacerbations one can decide on clinical grounds to prescribe a combination or one of anticholinergics and β_2-agonists.

Fig. 1. An excerpt from the guideline on COPD medication. Level 1 conclusions are supported strongly by literature. Remarkably this fragment is inconsistent with the intention of guidelines, as something is recommended for which there is no evidence.

The best indicator for lung function is FEV_1 (Forced Expiratory Volume in 1 second). It measures airway obstruction by testing whether the patient can overcome obstructive and restrictive resistance during forced exhalation. The resulting volume can then be compared with the predicted value based on physical properties such as age and gender. Higher values are thus an indication of better overall lung function. Percent values should be interpreted as a percentage of the expected value for someone with the same gender and age. Exacerbations, often caused by an infection or irritants from environmental air, result in a decrease in FEV_1 and worsening of other symptoms.

3.2 Temporal Logic

The guidelines express temporal dynamics (e.g. first prescribe drug A, if that fails prescribe drug B), which is cumbersome to describe in first-order predicate logic. In general, monitoring will result in temporal data, thus a representation over time is necessary. In order to simplify reasoning over time, we consider time as a discrete process consisting of an ordered set of points. We need to consider two different scenarios: decision-support based on monitoring data and on guideline execution. The data obtained from monitoring is time-indexed and thus allows analysing progression over time. Guidelines, on the other hand, use more abstract temporal concepts which cannot easily be represented with an exact time index without introducing a false sense of precision. Therefore our domain description uses time-indexed variables, with which it is possible to represent monitoring data. Our guideline formalisation will abstract from the time points and use temporal operators to describe general temporal relations.

The temporal logic operators we use can describe simple temporal relations but are limited in the level of temporal detail that can be expressed. The guidelines provide temporal information only occasionally, for instance the length of antibiotics treatment is given as 5-10 days, but the effect duration of bronchodilators is omitted completely as it depends on various patient specific factors. Hence slightly abstract temporal operators suffice to specify these relations intuitively. The temporal operators used are **H, G, P** and **F** [2], defined in Table 1.

Table 1. Definition of temporal operators

Notation	Informal meaning	Formal meaning
$\mathbf{H}\varphi$	φ has always been true in the past	$t \vDash \mathbf{H}\varphi \Leftrightarrow \forall t' < t : t' \vDash \varphi$
$\mathbf{G}\varphi$	φ will be true at all future times	$t \vDash \mathbf{G}\varphi \Leftrightarrow \forall t' > t : t' \vDash \varphi$
$\mathbf{P}\varphi$	φ is true somewhere in the past	$\vDash \mathbf{P}\varphi \Leftrightarrow \neg\mathbf{H}\neg\varphi$
$\mathbf{F}\varphi$	φ is true somewhere in the future	$\vDash \mathbf{F}\varphi \Leftrightarrow \neg\mathbf{G}\neg\varphi$

3.3 Formal Domain Description

For a decision-support system for COPD we need a clear description of the concepts involved. Structuring of concepts both clarifies the knowledge and enables richer descriptions, and hence an ontology form is useful. Tools such as Protégé/OWL [5], have been developed to construct domain ontologies, but the description logic used by OWL cannot express temporal relations very well. Therefore we use a first-order logic representation to define classes in a structured way, in Section 3.5 we will look at temporal extensions to description logic. Our domain description has a dual nature in the sense that we not only represent that a class can be a subclass of a more general class, but also that a superclass A consists of a specific number of subclasses $B_1, \ldots, B_m$. In logic this can be expressed by the bi-implication $\forall x (A(x) \Leftrightarrow (B_1(x) \vee \cdots \vee B_m(x)))$, meaning that instances of class A are instances of one or more classes $B_1, \ldots, B_m$, and that an instance of B_k is an instance of A. This representation, for example, permits us to infer that cough is a symptom but also that if a symptom is present is should be at least one of cough, dyspnea, fatigue, or sputum. As all formula concern patients, and all patient features occur at time points, we employ literals of the form $P(p, t)$, where t denotes time and p patients. In terms of class-subclass relationships, time t acts as the instance, similar to the variable x above, concerning a given patient.

In the domain of COPD this leads to the following classes, shown in Fig. 2. *Symptoms* are those things that a patient would report, *signs* can be measured by a clinician and *lab* results can be measured when lab equipment is available. *Conditions* are patient states that are not readily observable, *properties* are those features that are not strictly a symptom, sign or lab result. *Interventions* can be either *drug* or *non-drug* and some relevant environmental properties are given by the *environment* class.

In addition to these classes we need to specify the values that variables can take. Specifically we want to describe changes in patients over time, which we accomplish by statements like Cough(p, t,up), which for patient p at time t should be interpreted as an increase of coughing with respect to the previous time point. For the sake of brevity these specifications have been omitted, the intended meaning should be clear from context. In addition consistency requirements such as Cough(p, t,up) $\wedge$ Cough(p, t,down) $\vDash \bot$, stating that a symptom cannot get better and worse at the same time, should be added to the knowledge base.

$\forall p, t, x, y, z, w$ $\text{Symptom}(p, t)$ $\Leftrightarrow$
$(\text{Cough}(p, t, x) \lor \text{Dyspnea}(p, t, y) \lor$
$\text{Fatigue}(p, t, z) \lor \text{Sputum}(p, t, w))$

$\forall p, t, x, y, z$ $\text{Sign}(p, t)$ $\Leftrightarrow$
$(\text{FEV}_1(p, t, x) \lor \text{SaO}_2(p, t, y) \lor \text{Temp}(p, t, z))$

$\forall p, t, x, y, z$ $\text{Lab}(p, t)$ $\Leftrightarrow$
$(\text{PaO}_2(p, t, x) \lor \text{PaCO}_2(p, t, y) \lor \text{Sputum-sample}(p, t, z))$

$\forall p, t$ $\text{Condition}(p, t)$ $\Leftrightarrow$
$(\text{Airway-obstr}(p, t) \lor \text{Exacerbation}(p, t) \lor \text{Fit}(p, t) \lor$
$\text{Alveolar-hypovent}(p, t) \lor \text{Bact-inf}(p, t) \lor \text{Viral-inf}(p, t) \lor$
$\text{Hypoxemia}(p, t) \lor \text{Hypercapnia}(p, t))$

$\forall p, t, x, y, z$ $\text{Property}(p, t)$ $\Leftrightarrow$
$(\text{Activity}(p, t, x) \lor \text{BMI}(p, t, y) \lor \text{Breath-tech}(p, t) \lor$
$\text{Exacer-freq}(p, t, z) \lor \text{Smoking}(p, t))$

$\forall p, t$ $\text{Non-drug}(p, t)$ $\Leftrightarrow$
$(\text{Diet}(p, t) \lor \text{Oxygen-therapy}(p, t))$

$\forall p, t$ $\text{Drug}(p, t)$ $\Leftrightarrow$
$(\text{AB}(p, t) \lor \text{CS}(p, t) \lor \text{LAAC}(p, t) \lor$
$\text{LABA}(p, t) \lor \text{SAAC}(p, t) \lor \text{SABA}(p, t))$

$\forall p, t$ $\text{Environment}(p, t)$ $\Leftrightarrow$
$(\text{Cold-env}(p, t) \lor \text{Fog-env}(p, t) \lor \text{Smoke-env}(p, t))$

$\forall p, t$ $\text{Observable}(p, t)$ $\Leftrightarrow$
$(\text{Symptom}(p, t) \lor \text{Sign}(p, t) \lor \text{Lab}(p, t))$

$\forall p, t$ $\text{Intervention}(p, t)$ $\Leftrightarrow$
$(\text{Non-drug}(p, t) \lor \text{Drug}(p, t))$

Fig. 2. COPD formal domain description

3.4 Guideline Knowledge Base

With the classes we defined, we can describe patients and progressive patient status over time. Although useful for representing monitoring data it is too specific
to represent the information available in the guidelines. To formalise the useful
knowledge contained in the guidelines we will thus use the abstract temporal
operators defined above to describe temporal relations. A further problem with
the information in the guidelines is that no distinction is made between causal
knowledge about the world and recommendations. From a decision-theoretic
viewpoint the difference is important however. Causal knowledge tells us what
will change when we do something and from this we can infer what action gives
the desired result. The recommendations do not fit into this framework very
easily, we will get back to this point in Section 4. We will make the difference
explicit in the formalisation by superscripting the implication-symbol, for causal
relations ($\Rightarrow^c$) and recommendations ($\Rightarrow^r$), respectively.

To simplify notation in what follows, all statements are implicitly universally
quantified over patients and the patient variable p is left out of class predicates.
Furthermore we drop the time variable t as we cannot obtain values for it from
the guideline. The rest of this section describes the information in the guidelines
that forms the knowledge base contents.

Non-drug. The severity of COPD can be minimised with a number of best practises, either self-managed or with training. Quitting smoking is considered the most important and effective. It is here presented as a binary predicate, but a quantification in pack-years can be considered. The guidelines provide the following information, where BMI stands for Body Mass Index (kg/m^2); PaO_2 for arterial oxygen pressure (kPa); SaO_2 for arterial oxygen saturation. The first three formulae state causal consequences, the latter two state a clinical condition and their established treatment following the guidelines.

$\mathbf{H}$ Smoking $\wedge$ $\mathbf{G}$ $\neg$ Smoking $\Rightarrow^c$ $\mathbf{F}$ FEV_1(up)

$\mathbf{G}$ Activity(up) $\qquad\qquad\Rightarrow^c$ $\mathbf{G}$ Fit $\wedge$ $\mathbf{F}$ Dyspnea(down)

$\mathbf{G}$ Breath-tech $\qquad\qquad\Rightarrow^c$ $\mathbf{F}$ Dyspnea(down)

BMI < 21 $\qquad\qquad\qquad\Rightarrow^r$ Diet

$PaO_2 < 7.3 \vee SaO_2 < 90\%$ $\Rightarrow^r$ Oxygen-therapy

Drug. Before describing the guidelines for drug treatment an overview will be given of the intended effect of the prescribed drugs. The effects of antibiotics on lung function are unclear due to lack of recent relevant clinical literature. We will assume the general intended effect of antibiotics: combat bacterial infections. We will use the abbreviations SABA for Short-acting β_2-agonists; SAAC for Short-acting anticholinergics; LABA and LAAC are defined analogously for the long-acting variants; CS is short for Corticosteroids and AB for Antibiotics.

SABA $\Rightarrow^c$ $\mathbf{F}$ $(FEV_1$(up) $\wedge$ Dyspnea(down) $\wedge$ Fatigue(down))

SAAC $\Rightarrow^c$ $\mathbf{F}$ $(FEV_1$(up) $\wedge$ Activity(up) $\wedge$ Exacer-freq(down))

LABA $\Rightarrow^c$ $\mathbf{F}$ $(FEV_1$(up) $\wedge$ Exacer-freq(down))

LAAC $\Rightarrow^c$ $\mathbf{F}$ $(FEV_1$(up) $\wedge$ Dyspnea(down) $\wedge$

$\qquad\qquad$ Activity(up) $\wedge$ Exacer-freq(down))

CS $\quad\Rightarrow^c$ $\mathbf{F}$ Exacer-freq(down)

AB $\quad\Rightarrow^c$ $\mathbf{F}$ $\neg$ Bact-inf

Maintenance. These rules indicate under what circumstances which drugs should be prescribed for maintenance care according to the guidelines. Unfortunately the guidelines leave this mainly to the clinician, resulting in a somewhat unspecific formalisation.

$\mathbf{P}$ $\neg$ Symptom $\wedge$ Symptom $\Rightarrow^r$ SABA $\vee$ SAAC

$\mathbf{P}$ Symptom $\wedge$ Symptom $\quad\Rightarrow^r$ LABA $\vee$ LAAC

Exacer-freq(up) $\qquad\qquad\Rightarrow^r$ CS

Exacerbation. Exacerbations – episodes of worsening of health status – are defined in the guideline as an increase in dyspnea, sputum volume and purulence and an increase in cough. The first reaction to an exacerbation is administering a short-acting drug, if that fails to ameliorate the situation corticosteroids are prescribed. Furthermore given an exacerbation and fever or exceptional low lung function, antibiotics are prescribed.

Exacerbation $\qquad \Leftrightarrow^c$ Dyspnea(up) $\wedge$
$\qquad\qquad\qquad\qquad$ Sputum(up) $\wedge$ Cough(up)

Exacerbation $\qquad \Rightarrow^r$ SABA $\vee$ SAAC

Exacerbation $\wedge$ **P**(SABA $\vee$ SAAC) $\Rightarrow^r$ CS

Exacerbation $\wedge$
(Temp $> 38.5°\vee$ FEV$_1 < 30\%$) $\qquad \Rightarrow^r$ AB

Physiology. Although the guidelines give precise recommendations, underlying causes for health status lie outside the scope of the guidelines. Some additional background knowledge about pulmonary physiology can complement the knowledge from the guidelines and results in a more complete knowledge base. We incorporate information obtained through personal communication with a physiologist, which is also readily available from literature on the (patho-)physiology of the respiratory system. In order to not overcomplicate the KB however, we will refrain from including the many complex processes that are difficult to measure and whose relations to the knowledge from the guidelines is unclear. Although simplified, the addition of physiology gives some background information on how the respiratory system works. The following set of formulae contains the physiological processes and observable effects augmenting the guideline KB.

Bact-inf $\qquad\qquad \Rightarrow^c$ **F** Sputum-sample

Bact-inf $\vee$ Viral-inf $\qquad \Rightarrow^c$ **F** (Airway-obstr $\wedge$ Temp(up))

Airway-obstr $\qquad\qquad \Rightarrow^c$ **F** (Cough(up) $\wedge$ Sputum(up) $\wedge$
$\qquad$ PaO$_2$(down) $\wedge$ PaCO$_2$(up) $\wedge$ Alveolar-hypovent)

Alveolar-hypovent $\qquad \Rightarrow^c$ **F** (Hypoxemia $\wedge$ Hypercapnia)

Hypoxemia $\vee$ Environment $\Rightarrow^c$ **F** Dyspnea(up)

3.5 Description Logic

The kind of ontology information represented here is often formalised in description logic. The main advantage is that description logic is decidable, which makes it practically useful. Standard description logic does not provide temporal operators, but extensions exist that add tense logic operators similar to those we use here [6]. Although in general adding time increases the complexity it turns out that the language $\mathcal{CIQ}_\diamond$ is still decidable [7], and allows past and future operators.

From a practical point of view it is thus worth investigating whether our formalisation can be translated to description logic, which lets us inherit decidability. Description logic uses a class system for concepts which seems compatible with the classes we defined above. For example our Symptom-predicate can be translated to:

$$\text{Symptom} \equiv ((\text{Cough} \sqcap \exists\text{C-val.V}) \sqcup (\text{Dispnea} \sqcap \exists\text{D-val.V}) \sqcup$$
$$(\text{Fatigue} \sqcap \exists\text{F-val.V}) \sqcup (\text{Sputum} \sqcap \exists\text{S-val.V}))$$

A Symptom is defined as one of the concepts Cough, Dispnea, Fatigue or Sputum, each with a specific value. As ternary predicates are not allowed in description logic, separate classes are needed to specify the values each symptom can take, which is required to express changes in symptoms. In the example we only specified that a value exists, number restrictions can be used to restrict the possible values. This is equivalent to a first-order representation, in which the value can also be specified separately, for example:

$$\forall x \, \text{Symptom} \Leftrightarrow ((\text{Cough} \land \text{C-val}(x)) \lor$$
$$(\text{Dispnea} \land \text{D-val}(x)) \lor \cdots)$$

The temporal operators in both languages are the same (except for the symbols used). Since any formula in $\mathcal{CIQ}_\diamond$ can be quantified with a temporal operator, there is a straightforward translation from our temporal operators. So although we used a first-order representation here, it should be possible to translate this to description logic for an implementation that is decidable. Whether there is indeed a one-to-one mapping remains an open question.

4   Decision Theory

Now that we have constructed a knowledge base, the next step is reasoning with the knowledge. We first look at how we can use logical inference to find optimal decisions. As a second step we then add (qualitative) reasoning with uncertainty.

4.1   Inference

The temporal knowledge base (KB) from the guidelines can be used to make predictions about outcomes (OUT) based on observations (OBS) and interventions (INT). The optimal intervention can be chosen by inferring what the consequences of each intervention are for some choice of outcome. In this framework the recommendation rules cannot be used to make predictions because these rules only state a sort of black box decision. However, the casual rules from the guidelines, augmented with physiology, should in principle lead to the same optimal decision as the guideline recommendations. Therefore we can use the recommendations to verify the decision process.

Formally we can state the decision process as follows:

$$KB^c \cup \mathbf{P}\,\text{OBS} \cup \mathbf{P}\,\text{INT} \cup \mathbf{F}\,\text{INT} \vDash \mathbf{F}\,\text{OUT},$$

assuming that $KB^c \cup \mathbf{P}\,\text{OBS} \cup \mathbf{P}\,\text{INT} \cup \mathbf{F}\,\text{INT} \nvDash \bot$. The optimal decision is that future intervention that leads to the best outcome, with respect to some clinical criterion, given the causal knowledge base KB^c. We can then validate this decision-model result by comparing it to the recommended intervention. Assuming a sufficiently complete causal model and adding KB^r,

$$KB^r \cup KB^c \cup \mathbf{P}\,\text{OBS} \cup \mathbf{P}\,\text{INT} \vDash \mathbf{F}\,\text{INT},$$

should give the same intervention as the decision process.

As an interesting consequence of using class definitions, it is straightforward to reason about a particular subset of the knowledge base. For example we can write the meta-formula:

$$KB^c \cup \mathbf{P}\,\mathrm{Sign} \cup \mathbf{P}\,\mathrm{INT} \cup \mathbf{F}\,\mathrm{INT} \vDash \mathbf{F}\,\mathrm{Sign},$$

which expresses the notion of deriving the influence of interventions on signs only. The knowledge base now directly implies that only FEV_1, SaO_2 and Temp are relevant for this particular prediction.

4.2   Uncertainty

The representation in temporal logic provides an understandable format that allows making predictions based on observations. A problem with this representation however, is that uncertainty is not taken into account. The complexity of the disease and appropriate treatment cannot easily be put into the absolute terms the logic representation provides. Nonetheless the logic representation can serve as a basis for a more flexible decision network – a temporal Bayesian network augmented with decision nodes. The mapping is only partial however, for two reasons. Bayesian networks are generally propositional in nature, which means we lose expressive power compared to the temporal first-order logic. Second, inference in Bayesian networks requires probabilistic information which is not present in the guidelines and consequently also absent in the logic representation. At this point the best we can do is thus to build a qualitative decision network, which circumvents the second problem. In a later stage probabilistic information can still be added, by estimation or learned from data.

A probabilistic decision process models the uncertainty in the state and in the relations between the state and the observables. The actions influence how the state will change over time. In our domain this means the patient's status is partially observable and each intervention will affect with some probability how the status will change. In general this can be represented as a network of states influenced by the previous actions, where each state is partially observable (Fig. 3).

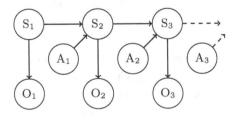

Fig. 3. Abstract representation of the layout of a decision network, where states are denoted by S, observables by O and actions by A. The indices denote time points.

Bayesian decision networks are a form of Bayesian networks with added decision nodes, which we can use to model our probabilistic decision process. In order to add information about uncertainty to the knowledge base, we need to relate the logic inference to the probabilistic decision networks, as follows:

$$\text{KB}^c \cup \mathbf{P}\,\text{OBS} \cup \mathbf{P}\,\text{INT} \vDash \mathbf{F}\,\text{OUT} \Rightarrow P(S_t \mid O_1, \ldots, O_{t-1}, A_1, \ldots, A_{t-1})$$

This means that the probability of a state at some time point t depends on previous states, of which we can only know the observed values, and actions up until time point t. Unfortunately, for a Bayesian network with many nodes it is computationally infeasible to take the complete history into account. We will assume however, that this problem is not too serious for our relatively small knowledge base.

We now have an abstract relation between the logic representation and a probabilistic representation, but in order to use the contents of the knowledge base in a probabilistic framework, we need probabilistic information. Some of the rules are deterministic in nature which results simply in a probability of 1 for the deterministic effect and 0 otherwise. Not all relations are so straightforward however, as uncertain effects are quite common but are not represented in the logic knowledge base. For these relations the probabilistic information will need to be estimated, either by hand with the help of a physician or, when relevant data is available, automatically with some parameter estimation technique (see e.g. [8] on Bayesian networks and decision graphs).

Fortunately, some probabilistic relations between state variables and observables can still be obtained when no data is available. Clinical instrumentation is well documented and the reliability is often reported in terms of sensitivity, specificity and positive predictive value, which give probabilistic information for the relation between measurement and true value. These reliability indicators are not quite sufficient however, because the probabilities needed in the decision network are conditional on actual observations. The information can nonetheless be used, because it provides constraints limiting the possible probability distributions.

Even when we have no probabilistic information, we can derive qualitative relations that are informative from the knowledge base. The representation in the knowledge base provides directions of change (e.g. Dyspnea(up)), which can be used as a qualitative indicator of the relation between variables (e.g. a smoky environment increases dyspnea). Similarly, the effect of drugs on state variables can be modelled qualitatively (e.g. CS decrease the probability of an exacerbation). These qualitative relations allow us to construct a qualitative probabilistic network [9] that codifies the logic information into a decision network. Qualitative relations have their expected meaning: a positive relation (+) between variables A and B indicates that a higher value of A makes higher values of B more likely; a negative relation (−) indicates that higher values of A make lower values of B more likely.

In Fig. 4 a part of a qualitative state definition is shown, with the corresponding observables for one time point. In our example we can measure the

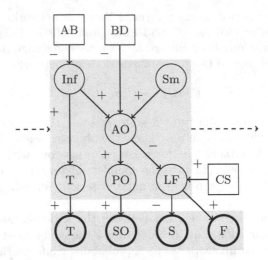

Fig. 4. COPD decision network. The grey square contains the decision network state-variables (Inf = Infection; Sm = Smoking; AO = Airway Obstruction; T = Temperature; PO = PaO$_2$; LF = Lung Function), the double circles in the bottom grey rectangle are the observables (T = Temperature; SO = SaO$_2$; S = Symptoms; F = FEV$_1$) and the rectangles depict the actions (i.e. drug prescription; AB = Antibiotics; BD = Bronchodilators; CS = Corticosteroids). The symptoms and bronchodilator drugs are represented as a single summary node for simplicity. The dashed arrows indicate the transitions between time points, assuming a repetitive (time invariant) network. The sign labels indicate qualitative relations between variables.

temperature, observe the symptoms and measure FEV$_1$ to assess lung function. The observation of symptoms can be quantified by asking patients to fill in standard medical questionnaires resulting in numeric scores that have well documented meaning. These observations can then be propagated through the hidden state network using propagation algorithms (which is possible both in quantitative [10] and qualitative [9] probabilistic networks). Given the observations, the effect of each possible action can be propagated to the next time step. The optimal action is the one leading to the state closest to the desired future state. For example given fever (indicating an infection) the optimal action will be administering antibiotics, decreasing the probability of infection at the next time step, which is represented by the negative relation in the decision network between nodes AB and Inf (Fig. 4).

Decision networks can be extended with explicit utility information indicating what conditions are desirable, which opens up the possibility of choosing the optimal action automatically by maximising utility. Deciding upon utilities however should be done in cooperation with a physician, as the consequences of incorrect utilities might be quite severe. In this paper we do not use utilities so we will not go into this any further.

An interesting open problem is the precise relation between the temporal logic operators and temporal decision-networks. Although the general meaning of 'somewhere in the future' is obvious, it is also very imprecise. This is a problem with the temporal granularity that can be derived from the guidelines, but specifically affects practical decision making. However, the temporal operators do convey some information that can be used probabilistically. The minimal consequence of for example, $\mathbf{F}$ Dyspnea(up), is that the probability of the patient's dyspnea improving is greater than 0. Additionally, assuming finite time and no uncertainty in the operators (i.e. $t \vDash \mathbf{F}x \Leftrightarrow \exists t' : t < t' < T : t' \vDash x$) the future (past) operator constrains the probability distribution. Although the shape of the distribution is not known, for $\mathbf{F}x$ the probability of x in the limit $t' \to T$ goes to 1. More realistic would be to allow uncertainty in whether x will happen, in which case the operator only increases the probability in some time frame after t. In future work it remains to further specify the relation between the probabilities in the decision network and the temporal logic operators.

5 Discussion

With a formal representation of the knowledge contained in medical guidelines, automated reasoning becomes possible. The quality of the decision-model depends on the quality of the guideline and the formalisation. Hence, evaluation of the model serves two purposes: checking the quality of the formalisation and verifying the guidelines. The former is necessary for any real-world application and should be based on patient data and clinical judgement, which lies outside the scope of the current work. Quality checking of guidelines has been studied before, for instance, checking diabetes guidelines [11]. The logic part of our representation could be used in a similar manner to verify properties of guidelines using a standard theorem prover. The temporal logic can be translated to a first-order logic representation that can be used as input for the theorem prover [11]. The advantage of the temporal logic representation is the clear interpretation and concise knowledge base, while the translation to first-order logic enables using powerful theorem provers.

Specific languages exist that provide various degrees of formalisation for medical guidelines (Asbru [12], PROforma [13], among others [14]). We have chosen temporal logic as it is both formal and expressive, although the latter can be considered a disadvantage compared to the more focussed alternatives. Also because the COPD guideline is not very specific on temporal aspects of treatments, the abstract temporal operators we use are more useful than Asbru's temporal patterns. However this will depend on the type of guideline, if a protocol with precise timing and duration information is available, Asbru's representation is an advantage. Since specific timing information is largely unavailable in the Dutch COPD guideline – which can be considered a shortcoming of the guideline – abstract temporal logic seems appropriate. A topic for future research is using probabilistic logic (see e.g. [15]), directly combining logic with probabilistic networks.

Building a clinically valid and practically useful decision-support system is a complex task that requires many steps. In this paper we have looked at formally representing knowledge that is available from medical guidelines, in particular in the domain of COPD. This has led to a two-step qualitative representation with temporal logic and qualitative probabilistic decision networks. Both representations have their strengths and weaknesses, which is why both have their use. A more in depth evaluation of usability still remains to be done, leading up to a COPD patient monitoring decision-support system.

References

1. Hommersom, A., Groot, P., Lucas, P., Balser, M., Schmitt, J.: Verification of medical guidelines using background knowledge in task networks. IEEE Transactions on Knowledge and Data Engineering (2007)
2. Lucas, P.: Quality checking of medical guidelines through logical abduction. In: Proceedings of AI-2003 Research and Development in Intelligent Systems XX (2003)
3. Viegi, G., Pistelli, F., Sherrill, D., Maio, S., Baldacci, S., Carrozzi, L.: Definition, epidemiology and natural history of COPD. Eur. Respir. J. (2007)
4. GOLD: Global Initiative for Obstructive Lung Disease, http://www.goldcopd.com
5. Knublauch, H., Ferguson, R., Noy, N., Musen, M.: The Protégé OWL plugin: An open development environment for semantic web applications. In: McIlraith, S.A., Plexousakis, D., van Harmelen, F. (eds.) ISWC 2004. LNCS, vol. 3298, pp. 229–243. Springer, Heidelberg (2004)
6. Artale, A., Franconi, E.: A survey of temporal extensions of description logics. Annals of Mathematics and Artificial Intelligence (2000)
7. Wolter, F., Zakharyaschev, M.: Temporalizing description logics. In: Proceedings of FroCoS 1998 (1998)
8. Jensen, F.V., Nielsen, T.D.: Bayesian Networks and Decision Graphs. Springer, Heidelberg (2007)
9. Wellman, M.: Fundamental concepts of qualitative probabilistic networks. Artificial Intelligence (1990)
10. Pearl, J.: Probabilistic reasoning in intelligent systems: networks of plausible inference. Morgan Kaufmann, San Francisco (1988)
11. Hommersom, A., Lucas, P., van Bommel, P.: Checking the quality of clinical guidelines using automated reasoning tools. Theory and Practice of Logic Programming (2008)
12. Miksch, S., Shahar, Y., Johnson, P.: Asbru: a task-specific, intention-based, and time-oriented language for representing skeletal plans. In: KEML 1997: 7th Workshop on Knowledge Engineering: Methods & Languages (1997)
13. Fox, J., Johns, N., Lyons, C., Rahmanzadeh, A., Thomson, R., Wilson, P.: PROforma: a general technology for clinical decision support systems. Computer Methods and Programs in Biomedicine (1997)
14. Terenziani, P., Molino, G., Torchio, M.: A modular approach for representing and executing clinical guidelines. Artificial Intelligence in Medicine (2001)
15. de Raedt, L., Kersting, K.: Probabilistic logic learning. SIGKDD Explorations Newsletter (2003)

Experiences in the Development of Electronic Care Plans for the Management of Comorbidities

Esther Lozano[1,*], Mar Marcos[2], Begoña Martínez-Salvador[2], Albert Alonso[1], and Josep Ramon Alonso[1]

[1] Hospital Clínic, Barcelona, Spain
[2] Universitat Jaume I, Castellón, Spain

Abstract. Recent studies have shown that care plans with comprehensive home interventions can be effective in the management of chronic patients. Evidence also exists about the importance of tailoring these care plans to patients, by integrating comorbidities. In this context, the development, implementation, outcome analysis, and reengineering of care plans adapted to particular patient groups earn relevance. We are concerned with the development and reengineering of electronic care plans dealing with comorbidities. Our hypothesis is that a library of reusable care plan components can facilitate these tasks. To confirm this hypothesis we have carried out an experiment consisting in developing a library of care plan components for the management of patients with COPD[1] or CHF[2], and next building a care plan for stable COPD&CHF patients by (re)using these components. In this paper we report on this experiment.

Keywords: Care plans, knowledge modelling, reusable components, guideline representation languages, PROforma.

1 Introduction

Recent studies have shown that care plans[3] with comprehensive home interventions can be effective in the management of chronic patients [2]. These plans include interventions that do not require a direct supervision by the physician, such as answering of questionnaires by the patient, or home visits and/or phone consultations by nurses (with access to the physician at the hospital for remote supervision). Most of these models, though, are exclusively targeted on one disease (i.e. COPD, CHF, Diabetes) in isolation. Evidence exists about the importance of integrating comorbidities in such models as a way of making them better

---

* This work has been supported by the Spanish Ministry of Science and Technology, through the research project TIN2006-15453 – HYGIA.

[1] Chronic obstructive pulmonary disease.

[2] Chronic heart failure.

[3] We use the term care plan to refer to a high level description of the overall plan of actions for the care of a patient, constructed from the basis of one or more guidelines that the patient should follow, similarly to the definition by Barretto [1].

D. Riaño et al. (Eds.): KR4HC 2009, LNAI 5943, pp. 113–123, 2010.
© Springer-Verlag Berlin Heidelberg 2010

tailored to patients. In such home-based health-care scenarios, the development and outcome analysis of care plans tailored to the management of particular patient groups earn relevance. Thus, if a concrete care plan does not prove to be effective according to predefined indicators it will have to be reengineered and implemented in its new form.

Several Artificial Intelligence (AI) tools can be used to facilitate the different phases of the process sketched above, generally speaking, care plan development, implementation, outcome analysis, and reengineering. The objectives of the HY-GIA project [3] are in line with this overall goal. In HYGIA we tackle the development of care plans in electronic format using existing AI languages for the representation of medical guidelines. Apart from serving as a basis for further development of dedicated decision support systems, the representation of care plans in these languages is important in terms of health-care knowledge capitalization.

Within the HYGIA project, we approach the development of care plans as a knowledge acquisition task during which the necessary knowledge is collected from different sources, mainly selected medical guidelines and experts' knowledge. At the same time, we concentrate on care plans for the treatment of co-morbidities. In this framework, our hypothesis is that a library of reusable care plan components can facilitate the development and reengineering of electronic care plans, even directly by medical experts [4].

To validate the above hypothesis, we have worked on a library of care plan components incorporating experts' knowledge about the necessary home interventions to manage patients under particular conditions, namely COPD or CHF. Then we have used these components to build a care plan for a specific group of patients, namely patients with both COPD and CHF who are stable, and with a clinical goal in mind, which is the prevention of exacerbations. In this paper we report on this experiment.

It is important to note that at this stage we do not aim at providing a fully functional tool to support care plan developers. Rather, this work should be considered as a "proof of concept" to validate our hypothesis, using existing guideline representation languages and tools. Consequently, the development of more specific tools with advanced features (e.g. web-based, collaborative) is beyond the scope of this work.

The structure of the paper is as follows. First section 2 details the experimental setting, including the modelling strategy that has been adopted and the guideline representation language (and tools) used. Then section 3 gives details on the care plan components that have been developed. Afterwards section 4 describes the development of the care plan for the management of COPD&CHF patients, including difficult issues and results. Finally section 5 presents our conclusions and points to the future work.

2   Experimental Framework

Our experiment consists in the development of a care plan in electronic format for the prevention of exacerbations in stable COPD&CHF patients, care plan which is partly based on home-based interventions. The need for such a specialised care plan, including the possibility of care plan reengineering, originates from ongoing clinical research at the Hospital Clínic of Barcelona [5], [2]. Likewise, the domiciliary interventions to be included in the care plan have been established by and are specific to this hospital. Finally, the recommendations for the diagnosis and treatment of COPD and CHF conditions are based on the recommendations from the guideline of the Global Initiative for Chronic Obstructive Lung Disease (GOLD) [6] and the guideline for the diagnosis and treatment of CHF of the European Society of Cardiology (ESC) [7], respectively.

The reason for choosing this specific patient population is that chronic respiratory diseases, particularly COPD, are an important and increasing burden of healthcare systems worldwide [5], [2]. COPD exacerbations, mostly occurring in elderly people with concurrent chronic comorbidities such as CHF, often generate dramatic increases in admissions to hospital emergency services, with the subsequent negative impact on patient quality of life, prognosis and costs.

With regard to the care plan itself, note that we entirely rely on the expertise of researchers of Hospital Clínic. Thus, all the medically relevant aspects (e.g. severity of the disease) have been specified in the care plan by the hospital experts. Besides, issues like the clinical validation of the care plan using cohort data is considered to be outside the scope of this work. Instead we focus on the development aspects from a knowledge engineering perspective, with the goal of facilitating care plan reengineering.

In this light we have identified two important issues related to the experimental framework. First, it must be clarified to which extent the care plan knowledge is based on the experts' knowledge and the guidelines, to determine the acquisition and subsequent modelling strategies. Second, it is important to decide on the appropriate guideline representation language and tools, practicable for the purposes of care plan development and reengineering from components. The following subsections deal with these issues.

2.1   Knowledge Acquisition Framework

As mentioned before, the knowledge necessary for the development of the care plan has both the hospital experts and the GOLD and ESC guidelines as sources. With the purpose of determining the scope of each of these knowledge sources, initial interviews were conducted. As result, it became clear that the care plan is structured into three main phases, namely *admission, re-evaluation* and *follow-up*, each of them including interventions for both COPD and CHF.

The *admission* phase deals with the initial assessment of the disease and determination of the treatment, as well as with the complementary interventions performed in the Hospital Clínic (mainly completion of questionnaires) to determine if the patient can be enrolled in the care plan. The knowledge for the interventions in this phase comes mainly from the guidelines, in the case of the assessment and the determination of the treatment, but also from the hospital experts, in the case of the complementary actions. The *re-evaluation* is a short phase carried out at the patient's home. In this phase, the data provided by the patient in the questionnaires (e.g. about his/her knowledge of the disease) are revised, and educational directions are given based on the weak points detected. Finally, in the *follow-up* phase the status of the patient is periodically assessed through anamnesis and clinical examination. In case a critical condition is detected, such as cardiac decompensation, the patient should leave the care plan and enroll in a more appropriate one. Another important intervention in this phase is the revision of the treatment, if needed. The knowledge for the re-evaluation and follow-up phases and comes mainly from the experts, specially in the case of interventions for the management of comorbidities.

Therefore, we are in a context in which guideline knowledge must be combined with experts' knowledge in a care plan structured according to the three phases described above, which bears little relation to the GOLD and ESC guidelines. In such a context, we have opted for modelling components to be used as care plan building blocks, instead of modelling the guidelines and then selecting and adapting parts thereof. The former approach can be named as *"care plan oriented"*, as opposed to the latter *"guideline oriented"* one. Figure 1 shows a schematic view of the *"care plan oriented"* approach we have followed.

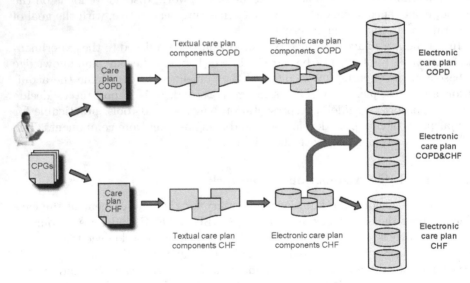

Fig. 1. Strategy for the development of a care plan for COPD&CHF management

With a *"care plan oriented"* approach the acquisition&modelling strategy follows a path like:

textual care plan (with expert and guideline knowledge) → textual care plan components → electronic care plan components → electronic care plan instead of:

textual guidelines → *electronic guidelines* → *electronic care plan (with expert knowledge)*

The distinctive features of the *"care plan oriented"* approach are (1) the central role of the knowledge needed for the care plan, (2) the focus on care plan building blocks (or components), and (3) the point of the process at which the specific expert knowledge is fed, which is earlier than in the case of the *"guideline oriented"* approach.

Lastly, and with the purpose of facilitating the development task (while increasing the possibilities of reuse of the care plan components in the library), we have opted for first modelling the care plan components for the management of COPD and CHF separately (both for stable patients), then building the respective care plans, and finally developing the care plan for the management of COPD&CHF patients. Figure 1 also illustrates the creation of separate COPD and CHF care plans prior to the development of the overall one.

2.2   Guideline Representation Language

In addition to the acquisition&modelling strategy, the choice of the knowledge representation language is another important aspect of our experiment. Within the HYGIA project we have committed ourselves to the use of one of the existing guideline representation languages. Among the leading languages for guideline representation cited in recent reviews [8], [9], [10], we examined Asbru, GLIF, PROforma, and SAGE. The selection was based on a set of requirements, mainly the availability of a modelling tool with graphical elements, the availability of a suitable execution engine, and the adequacy of the modelling&execution tools for the development of both care plan components and care plans of some size (this includes e.g. the possibility of nesting of tasks). The PROforma language was chosen based on these requirements, and also taking into consideration the expertise of the participants in the HYGIA project.

The PROforma language was developed at the Imperial Cancer Research Fund by Fox and colleagues [9], [11]. In PROforma a guideline is modelled in terms of tasks hierarchically organised in plans. PROforma tasks fall into four categories, namely: actions, enquiries, decisions and plans. Plans are used to group together other tasks e.g. so that they are performed at the same time. PROforma processes can be represented as directed graphs in which different types of nodes represent tasks and arcs represent scheduling constraints. Tasks may have a number of properties that determine the way they must be executed. For instance, plans can have preconditions and task scheduling constraints (these properties are shared with the rest of task categories), but also termination and abort conditions.

With respect to the tools, we have used the Tallis toolset [12]. Tallis includes two applications: the *composer*, which is a graphical environment for the creation and editing of guidelines, and the *tester*, which is an integrated tool for enacting guidelines step by step. Sections 3 and 4 below describe the components and the care plans that have been developed as part of our experiment. These sections also include some examples of PROforma processes.

3 Care Plan Components for the Management of COPD and CHF

As explained previously, we have first developed the care plan components for the management of COPD and CHF. This section includes examples taken from the COPD care plan. Figure 2 contains a screenshot of the Tallis tool showing the task network representing the *COPD_management* care plan. This network includes three plan nodes (represented as rounded boxes), one for each of the main phases described in section 2, together with the scheduling constraints among them (represented as directed arcs). In this case, the constraints impose a sequential control flow among the plans. Moreover, in figure 2 we can see a hierarchical decomposition of the PROforma tasks used in the description of the COPD care plan, among other things.

Figure 3 shows the task network pertaining to the *COPD_diagnosis* plan, which is part of the *COPD_admission* one. In this case the network includes enquiry nodes (represented as diamonds) to gather the necessary clinical data, and decision nodes (represented as circles) to subsequently confirm the COPD diagnosis and determine the severity of the disease. The latter has been modelled by means of action nodes (represented as boxes).

We posit that the PROforma plans in figure 2, as well as other plans at lower levels such as *COPD_diagnosis* (the details of which are shown in figure 3), constitute the components from which more complex care plans related to the COPD condition can be constructed.

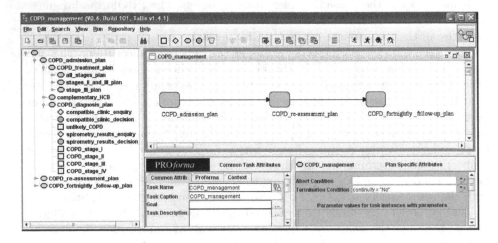

Fig. 2. PROforma model of a care plan for COPD management

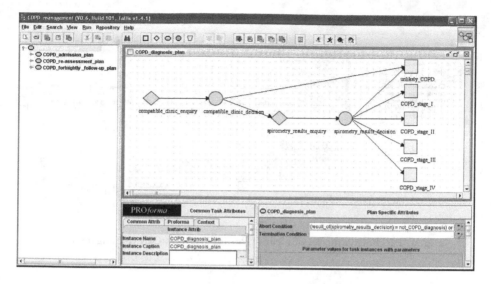

Fig. 3. PROforma model of COPD diagnosis

4 Care Plan for the Management of Stable COPD&CHF Patients

Based on the care plan components developed in the first place, we have constructed a care plan for the management of COPD&CHF patients. At a high level, the care plan also consists of three main plans, namely admission, re-evaluation and follow-up. Each of these plans has been developed by reusing different parts of the COPD and CHF plans, mostly whole plans but also different plan elements. The example below tries to illustrate this.

The PROforma network in figure 4 represents the plan *COPD − CHF_management*. This plan not only has the same structure as the plan in figure 2, but also contains the same definition for certain tasks (e.g. *COPD_diagnosis* and *COPD_treatment*, see the hierarchical decomposition in figure 4).

We have identified several scenarios regarding the reuse of care plan components. In some cases, e.g. in the case of the diagnosis and treatment plans, the PROforma plans have been directly reused (i.e. cut&pasted) as off-the-shelf components. This implies that both the layout of the task network and the tasks themselves are reused. Otherwise, when direct reuse was not possible or not deemed appropriate for some reason, different solutions have been adopted. The reuse scenarios that we have identified are the following:

1. direct reuse of plan (task layout plus tasks) as an off-the-shelf component.
2. reuse of plan (task layout plus tasks) with small changes, either (a) minor changes in tasks, e.g. changes in task preconditions, or (b) addition of a small number of tasks. In this scenario the original layout and tasks are preserved.

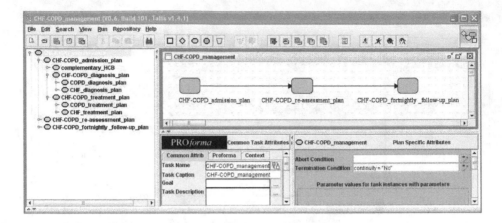

Fig. 4. PROforma model of a care plan for COPD&CHF management

3. reuse of task layout with replacement of tasks by similar ones, mostly different versions thereof. Note that this scenario also covers the cases in which the resulting plan includes additional elements, provided that the original task layout is preserved.

In addition to direct reuse, note that the reuse of plans with small changes and the reuse of task layout with replacement of tasks (i.e. scenarios 2 and 3) in general do not require great effort and produce actual modelling benefits. The gain of reuse depends on the complexity and size of the object reused. To give an idea of the latter, the plans reused in our experiment contained 3 to 40 tasks.

As example of situation where the modelling effort was higher we can mention the adjustment made to the plan *class_I_plan* (therapy plan within *CHF_treatment_plan* for class I patients) so that it could be used in the COPD&CHF care plan. Concretely, the part for the (CHF) beta-blocker therapy had to be adapted to cope with contraindications that apply to a subset of COPD patients. Although this example falls in scenario 2(a), it required further acquisition of specialised knowledge on comorbidities and hence it was considered harder. This has led us to the conclusion that plans should be documented with contraindications of the procedures they contain, to facilitate the reuse/adaptation in these situations.

Table 1 shows some numbers from our modelling experiment. The table lists the number of occurrences of the different reuse scenarios, together with the total number of (sub)tasks involved in the reuse (note that reusing a whole plan implies reusing all its subtasks) and the percentages with respect to the total number of tasks in the care plan. In this total we have disregarded duplicate plans (which would have been copied&pasted anyway, with or without component reuse), resulting in 136 subtasks. Likewise, we are not looking at duplicate parts to count reuse occurrences. Finally, note that in some cases the number of involved subtaks is approximate, due to the nature of the reuse scenario.

Table 1. Reuse of tasks in the development of a care plan for COPD&CHF management from components

Reuse scenario	Number of occurrences	Number of (sub)tasks reused (and % w.r.t. total[a])
scenario 1, reuse of plan	3	64 (47%)
scenario 2, reuse of (a) plan w/small changes (b)	1 2	41 (30.1%) 19 (14%)
scenario 3, reuse of task layout	3	N/A[b]

[a] The total number of tasks (excluding duplicate plans) is 136.
[b] Reuse of task layout relies on new subtasks.

In terms of the number of subtasks reused, the most important reuse scenarios are 1 and 2(a), i.e. direct reuse and reuse of plans with minor changes in tasks. All in all, mentioning that although other types of reuse are not so important in terms of subtasks reused, they occur in practice and prove to be important in care plan development (see e.g. the reuse of layout with task replacement in figure 4). Finally, pointing out that nearly all the plans are more or less based on previous components (over 90% of subtasks). Therefore, we can consider that the library of care plan components has simplified the construction of the COPD&CHF care plan to a great extent.

5   Conclusions

Recent studies have shown that care plans with comprehensive home interventions can be effective in the management of chronic patients. Evidence also exists about the importance of tailoring these care plans to patients, by integrating comorbidities. In this context, the development, implementation, outcome analysis, and reengineering of care plans adapted to particular patient groups earn relevance.

We are concerned with the development and reengineering of electronic care plans dealing with comorbidities. Our hypothesis is that a library of reusable care plan components can facilitate these tasks. To confirm this hypothesis we have carried out an experiment consisting in developing a library of care plan components for the management of patients with COPD or CHF, and next building a care plan for stable COPD&CHF patients by (re)using these components. For this purpose we have used the PROforma guideline representation language and its associated tools.

We have succeeded in building a complete care plan for a practical application from the care plan components previously developed, as shown in section 4. This has been done by either directly reusing the components as they were, or by reusing parts of the description of these elements, e.g. the task layout. We consider that the reuse of care plan components has simplified the construction of the care plan to a great extent. On the downside we should mention that there are situations which require additional acquisition&modelling work, even with component reuse. Still, this additional work can be preserved in the library, in the form of a new, more specialised care plan component. Section 4 includes an example of such situations, which is related to the fact that COPD and CHF are comorbidities.

The *"care plan oriented"* approach, focused on care plan modelling rather than on guideline modelling, has proven to be very useful at least in two respects. First, we have limited ourselves to modelling those guideline parts which were required for the care plan, disregarding the rest. E.g. the parts dedicated to the management of very severe COPD patients have not been modelled, because they are outside the scope of the care plan. And second, the components are designed for use in care plans for home-based management of patients and thus incorporate knowledge not covered in the guidelines. As result, the components incorporate specific knowledge that otherwise would have to be added in later stages.

With respect to the language and tools, we can say that they are well adapted to our purposes. The PROforma language has served to describe the care plan knowledge in a convenient manner. On the negative side we can mention the lack of language elements to describe concepts which are needed in care plans, such as roles and actors. As for the tools, we can highlight as advantages the graphical editing as well as the integration with an execution engine (the tester), which facilitates care plan debugging. An aspect where improvements could be made is the support for cut&paste of tasks, which is crucial for our purposes. A better support is to be expected in the commercial version of PROforma tools.

As future work we plan to carry out the same exercise with different combinations of diseases, building care plans to deal with COPD, CHF, and/or Diabetes. With a wider case study we will be better placed to determine e.g. how the library components must be documented to facilitate reuse/adaptation (e.g. with contraindications of the procedures within a plan). In parallel, we will explore other solutions to the representation of electronic care plans, possibly different from guideline languages.

In addition to the above, we will study the suitability of the framework for the development and reengineering of electronic care plans directly by care plan developers. Since the languages and tools we use come from the AI in Medicine field, it is crucial to ensure that they can be used by medical experts with no effort. Related to this issue and to the usability of electronic care plans developed, we plan to put into operation a decision support system to monitor a set of indicators as defined by care plan developers. In case these indicators show that the care plan is not appropriate for the target patient group, it will have to be reengineered and integrated into the decision support system in its new form.

References

1. Barretto, S.A.: Designing Guideline-based Workflow-integrated Electronic Health Records. PhD thesis, School of Computer and Information Science, University of South Australia (2005)
2. Casas, A., Troosters, T., Garcia-Aymerich, J., Roca, J., Hernández, C., Alonso, A., del Pozo, F., de Toledo, P., Antó, J., Rodríguez-Roisín, R., Decramer, M.: Integrated care prevents hospitalisations for exacerbations in COPD patients. European Respiratory Journal 28, 123–130 (2006)
3. HYGIA Project, http://banzai-deim.urv.net/~riano/TIN2006-15453/ (Date of access: May 2009)
4. Alonso, A., Marcos, M., Alonso, J., Gelabert, G., Martínez-Salvador, B., Riaño, D., Taboada, M.: A knowledge-acquisition framework to facilitate the development and reengineering of care plans in electronic format. In: Proc. of the Tromsø Telemedicine and eHealth Conference, TTeC-2008 (June 2008)
5. Hernandez, C., Casas, A., Escarrabill, J., Alonso, J., Puig-Junoy, J., Farrero, E., Vilagut, G., Collvinent, B., Rodriguez-Roisin, R., Roca, J.: Partners of the CHRONIC project: Home hospitalisation of exacerbated chronic obstructive pulmonary disease patients. European Respiratory Journal 21, 58–67 (2003)
6. Global initiative for chronic Obstructive Lung Disease: Global Strategy for the Diagnosis, Management and Prevention of Chronic Obstructive Pulmonary Disease. Executive Summary (2006), http://www.goldcopd.com/
7. The Task Force for the Diagnosis and Treatment of Chronic Heart Failure of the European Society of Cardiology: Guidelines for the diagnosis and treatment of chronic heart failure: executive summary (update 2005). European Heart Journal 26, 1115–1140 (2005)
8. Peleg, M., Miksch, S., Seyfang, A., Bury, J., Ciccarese, P., Fox, J., Greenes, R., Hall, R., Johnson, P., Jones, N., Kumar, A., Quaglini, S., Shortliffe, E., Stefanelli, M.: Comparing computer-interpretable guideline models: A case-study approach. Journal of the American Medical Informatics Association 10, 52–68 (2003)
9. de Clercq, P., Blom, J., Korsten, H., Hasman, A.: Approaches for creating computer-interpretable guidelines that facilitate decision support. Artificial Intelligence in Medicine 31, 1–27 (2004)
10. Isern, D., Moreno, A.: Computer-based execution of clinical guidelines: a review. International Journal of Medical Informatics 77(12), 787–808 (2008)
11. Sutton, D., Fox, J.: The syntax and semantics of the PROforma guideline modeling language. Journal of the American Medical Informatics Association 10(5), 433–443 (2003)
12. Tallis Toolset, http://www.cossac.org/tallis/index.html (Date of access: May 2009)

Challenges in Delivering Decision Support Systems: The MATE Experience

Dionisio Acosta[1], Vivek Patkar[1], Mo Keshtgar[2], and John Fox[3],*

[1] Cancer Institute, University College London, UK
[2] Department of Surgery, Royal Free Hospital, London, UK
[3] Department of Engineering Science, University of Oxford, Parks Road, Oxford,
OX1 3PJ, UK
john.fox@eng.ox.ac.uk

Abstract. Cancer Multidisciplinary Meeting (MDM) is a widely endorsed mechanism for ensuring high quality evidence-based health care. However, there are shortcomings that could ultimately result in unintended patient harm. On the other hand, clinical guidelines and clinical decision support systems (DSS) have been shown to improve decision-making in various measures. Nevertheless, their clinical use requires seamlessly interoperation with the existing electronic health record (EHR) platform to avoid the detrimental effects that duplication of data and work has in the quality of care. The aim of this work is to propose a computational framework to provide a clinical guideline-based DSS for breast cancer MDM. We discuss a range of design and implementation issues related to knowledge representation and clinical service delivery of the system, and propose a service oriented architecture based on the HL7 EHR functional model. The main result is the DSS named MATE (Multidisciplinary Assistant and Treatment sElector), which demonstrates that decision support can be effectively deployed in a real clinical setting and suggest that the technology could be generalised to other cancer MDMs.

Keywords: Cancer Multidisciplinary Meeting, Decision Support System, Clinical Guideline, HL7 Functional Model, Electronic Health Record.

1 Introduction

The complex nature of cancer poses significant risks of poor coordination, miscommunication and unintended patient harm. The cancer Multi Disciplinary Meeting (MDM) is widely endorsed as a mechanism for ensuring quality of care. MDMs are designated weekly meetings involving all core members of the care team, where all cancer patients are reviewed and management decisions are taken collectively. However, there are shortcomings in the current conduct of MDMs that have made it a high priority for the National Health Service (NHS) in the UK. These include inadequate documentation of both patient data and MDM decisions, missed opportunities to screen patients suitable for research trials and

---

* Corresponding Author.

D. Riaño et al. (Eds.): KR4HC 2009, LNAI 5943, pp. 124–140, 2010.
© Springer-Verlag Berlin Heidelberg 2010

lack of appropriate tools for auditing and monitoring the MDM performance. Clinical guidelines and Clinical Decision Support (CDS) technology have been shown to improve decision-making on various measures, and improvements in the NHS information technology infrastructure provide an opportunity to roll out this technology on a significant scale. In this paper we describe the design, implementation and deployment of a decision support system (DSS) for supporting the Breast Cancer Multidisciplinary Meeting (MDM) which is currently being trialled at the Royal Free Hospital, London UK.

MATE is a PRO*forma*-based decision support system that provides the clinical team with a computational tool in which patient investigations are recorded and specific recommendations given in real time during the MDM. MATE is based on a consensus framework from 16 breast cancer clinical guidelines. The ultimate aim of the system is to integrate seamlessly with existing and forthcoming hospital information technology infrastructure, from patient electronic records to PACS, avoiding the detrimental effects that duplication of efforts on behalf of hospital staff has in adopting new technologies. This aim cannot be achieved in a single step given the very particular requirements laid upon systems in a clinical setting, hence the need for a clear road-map that would steer the design and development work.

We adopted the HL7 Electronic Health Record System Functional Model (EHR-S FM) for this purpose, as a framework to drive the requirements discussions and motivate design decisions for CDS services. This approach has an intrinsic benefit that it provides a comprehensive set of system requirements, which are otherwise impossible to collect in the context of enterprises such as MATE. On the other hand, it has the disadvantage that it could increase the complexity of the design to an extent that may become a burden and jeopardise the primary aim of deploying the DSS.

In this paper we describe MATE and discuss a range of design and implementation issues, particularly those regarding knowledge representation and clinical service delivery. We will summarise EHR-S FM and address how the requirements derived from EHR-S FM affect the design and implementation of the DSS that we expect will facilitate portability and interoperability of the application. We will also discuss general implementation and deployment issues for a service-based architecture for DSS.

The paper is organised in four sections. The methodological aspects regarding knowledge modelling and decision making, data modelling, system architecture and interoperability are outlined in Section 2. Section 3 describes and discusses the results obtained in terms of the decision support system and the proposed service oriented interoperability architecture. In particular, this section contrasts this work against others found in the literature. Finally, concluding remarks are presented and further work is proposed in Section 4.

2   Methodology

Three areas needed to be addressed during the course of this work, namely knowledge representation and decision making, data modelling and finally interoperability. The next sections describe the methodological considerations pertaining to each one of these.

2.1   Knowledge Modelling and Decision Making

In the field of breast cancer, knowledge is given by a compendium of clinical guidelines provided in text form. Each one describes particular patient presentations, recommends the actions to be taken and provides references to the underpinning evidence. Some of these guidelines cover disjoint areas of treatment while others overlap to a certain extent. This requires a knowledge modelling methodology that not only accounts for simple cause/effect interactions, but also would allow us to fuse different knowledge sources in order to compute treatment recommendations. Furthermore, this framework should effectively convey the supporting evidence to the medical staff, i.e. the arguments for and against and references to the corresponding studies (observational studies and clinical trials).

PRO*forma* [1] provides a task class hierarchy and an argumentation-based decision making framework that satisfies the requirements specified above. Figure 1 depicts the PRO*forma* class hierarchy, with which clinical guidelines can be modelled as a set of decisions, each one is comprised of several non-exclusive candidate options, which in turn are promoted or demoted by a set of arguments. In our case, arguments are truth-valued expressions that model the patient presentation for a particular clinical evidence scenario. For additional information on PRO*forma*, its associated *Tallis* tool-set and evaluations of its modelling capabilities, please refer to [1,2].

The rest of this section present the methodological aspects considered when modelling decisions in the PRO*forma* language, in particular the heuristic used

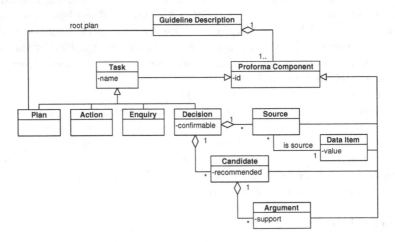

Fig. 1. UML class diagram of the PRO*forma* class hierarchy adapted from Sutton et al. [3], depicting the class attributes relevant to the breast cancer MDM guidelines

for selecting argument weights. In addition, they describe the adopted controlled terminology and the knowledge base verification process.

Decisions, Candidates and Arguments. Following the PRO*forma* framework, a breast surgeon (VP) extracted decisions, candidates and arguments from a compendium of 16 guidelines [4] and modelled them using the *Tallis* tool-set. Momentarily ignoring breast laterality and lesion multiplicity, the PRO*forma* breast cancer MDM knowledge representation considers 10 decisions, modelled using 61 candidates, 220 arguments and 96 data items. In addition, breast laterality and lesion multiplicity were also taken into consideration, allowing the system to provide specific recommendations.

The output of this process is a single PRO*forma* representation of the guideline compendium, where all treatment recommendations –decision candidates– are modelled. As some of these overlap between different guidelines, the corresponding arguments are harmonised using the argument weight mechanism provided by the PRO*forma*language and also rewriting the logic expressions whenever possible. It is important to note that this process ensures that arguments can always be traced back to their corresponding guidelines and support evidence given in terms of studies.

Controlled Terminology. With the aim of facilitating semantic interoperability, data items used for guideline modelling followed a controlled terminology standard. For each data item required, a manual search using the NCI CDE Browser [5] was performed, obtaining the public id and permissible values whenever a direct match was possible. Alternatively, a controlled term was derived by decomposing the original one and its set of values into its constituent words and looking for these in the NCI CDE repository. Furthermore, a default *unknown* value was added to each data item, in order to account for unavailable data in the knowledge model.

Argument Weight Selection Heuristic. An important aspect of the knowledge modelling framework in our context, is the ability to fuse the evidence supporting/demoting a decision candidate. As decision candidates overlap between different guidelines, the corresponding arguments need to be harmonised to produced a single computable form and for this purpose we employ arguments weights.

PRO*forma* arguments have associated weights which could be either symbolic – for or against – or numeric with no range restrictions. As there is no given methodology for assigning argument weights in the PRO*forma* framework, we propose an heuristic for this purpose.

Following the work by the GRADE working group [6], our heuristic estimates the argument weight w as the product between the recommendation strength s and the evidence quality q, as shown in Table 1. We complemented the original grading system by adding an overriding recommendation strength for treatment contraindications.

Table 1. Recommendation grading system adapted from Guyatt et al. [6] in matrix form, where rows correspond to evidence quality (q) and columns to recommendation strength (s). The entries describe the implications and give the argument weight as $w = q \times s$.

Quality (q) vs. Strength (s)	Contraindication ($s = 10^2$)	Benefits $\gg$ Risks ($s = 10^1$)	Benefits $\sim$ Risks ($s = 10^0$)
RCTs or overwhelming observational evidence ($q = 1.0$)	never apply ($w = 10^2$)	apply without reservation ($w = 10^1$)	weak recommendation ($w = 10^0$)
RCTs with limitations or strong observational evidence ($q = 0.5$)	never apply ($w = \frac{1}{2}10^2$)	apply without reservation ($w = \frac{1}{2}10^1$)	weak recommendation ($w = \frac{1}{2}10^0$)
Observational evidence or case series ($q = 0.1$)	never apply ($w = 10^1$)	strong recommendation ($w = 10^0$)	alternatives reasonable ($w = 10^{-1}$)

The heuristic assigns numerical values to the qualitative recommendation strength by mapping it to an exponential decay, modelling our expectation on how trust in the recommendation diminishes as the balance between benefits and risks is reached. Similarly, the quality value is mapped to a linearly decreasing function. The rationale of this mapping is that arguments with higher quality evidence but low strength should not overpower arguments with high strength but low quality, for instance in the case of contraindications.

Decision Workflow Modelling. The nature of the MDM requires all decisions to be available (confirmable) to the user at the same time, although there are clear precedences among them. Therefore, we resort to set them all as acting in parallel fashion and modelled precedences using data dependencies (wait conditions in the PRO*forma* language). This PRO*forma* task sequencing mechanism ensures that arguments, candidates and decisions are re-evaluated interactively as required by the user, while enforcing precedences.

In addition to the above, a series of automatic tasks were included in the knowledge base for prognostications and alerts. For example, T, M and N staging investigations are automatically suggested[1], as are Adjuvant! Online style prognostications [7], adapted for British mortality rates, and MSKCC nomogram [8]. It is interesting to note, that once a structured model has been built, a greater level of automation can be easily achieved, providing complete investigations to the MDM staff.

---

[1] T, N, M staging is an internationally accepted staging system to uniformly categorise the extent of cancer in a patient and prognosis. T, N and M stands for Tumour size, Nodal involvement and Metastatic spread respectively.

Knowledge Base Verification. In order to verify the correctness of the system recommendations, we employed a black-box approach in the form of an audit study [9]. During the audit, the system was run mirroring the MDM sessions for a period of time and all discrepancies between the system recommendations and the MDM were recorded, analysed and fed back into the system. In all instances the discrepancies found could be stratified into programming errors or legitimate non-compliance. The former were corrected, not requiring adjustment of the decision model. The latter represented around 6% in case of disagreement with guidelines and up to 60% in the case of missed opportunities for identifying potentially eligible patients for clinical trials. The weekly test units were run for a period of time until a sufficiently representative sample of MDM cases were collected and analysed (400 cases in a six months period). This methodology provided empirical evidence that the system recommendations were in line with the guideline and MDM recommendations. Although we are aware of other techniques that could have been employed for facilitating this task, such as guideline critiquing [10,11], automation of these is guideline-representation dependent and thus not readily applicable in our context.

2.2 Data Modelling

Whether a DSS is either a service or an integral component of the EHR-S, it requires a persistence component for its decision support related data and underlying evidence, primarily to provide a process time-line, guideline version control and for audit functions. On the other hand, there is no explicit contract between the EHR-S and the DSS that would ensure that data required for decision support data is readily available. The data might not be modelled within the EHR-S or could be present in a format which renders it unusable. Furthermore, it should

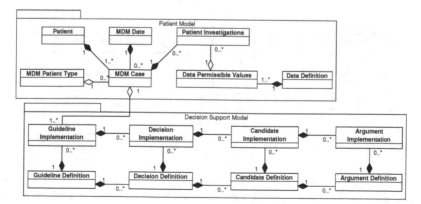

Fig. 2. UML class diagram of MATE data model. The Figure depicts the patient and decision support data models. The decision data model supports the prospective and retrospective application of different guidelines to every MDM case.

be possible to re-run the decision support process for audit purposes and for this reason data could be stored in a DSS patient data model.

Following these considerations, the DSS data model consists of two main components: the patient proxy data model, which contains the patient's static knowledge, and the decision data model (see Figure 2). Each patient investigation is controlled by a data permissible values enumeration entity. This enumeration entity models the controlled vocabulary used and its key references the coding scheme designator, code name, code value and code meaning. The first three map the data value to the exact controlled term, whilst the meaning provides a human readable interpretation of the code.

The decision support data model mimics the PRO*forma* class hierarchy model for decisions, candidates and arguments, but does not consider workflow. The Model provides persistence to guidelines both at design and at run-time, using definition and implementation sub-models respectively. The former replicates the PRO*forma* class hierarchy whilst the latter models the information generated during the guideline enactment, in particular arguments truth value, candidates recommended and confirmed decisions, for each patient and MDM session.

2.3   System Architecture

The system architecture follows the standard Model-View-Controller (MVC) design pattern [12]. The View is implemented by a user interface which emphasises direct access to every patient data item without the need for browsing. The Model implements the patient model, and the decision support data model as described in detail in Section 2.2. The Controller coordinates the interactions between the user interface and the model, in particular it initiates the changes to the model required when decision support is requested by the user. Figure 3 depicts the system architecture identifying in each case the underlying technology.

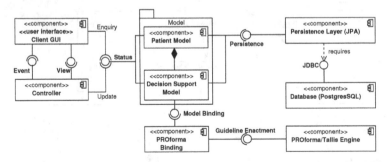

Fig. 3. System architecture UML component diagram depicting the Model-View-Controller pattern. Decision support interfaces with the *Tallis* engine via a binding component, to perform updates which are immediately notified to the View (user interface).

2.4   Decision Support and Interoperability

In order to gather and analyse interoperability requirements, we considered the HL7 EHR-S FM [13], and adjusted the design and implementation of the system accordingly. HL7 EHR-S FM defines a standardised set of functions that might be present in an EHR-S, in the form of a conceptual description given from the user perspective in a user-oriented language. Figure 4 shows the first level functions considered in the model. This set of functions covers all those considered essential in at least one health-care application and deals with interoperability from the semantical and technical points of view. The model neither makes assumptions on the specific architecture or contents of the EHR-S, nor endorses any given technology.

The Clinical Decision Support functions considered in the model are themselves classified into 7 functional groups as follows:

Manage Health Information support for standard and context driven patient assessments, detection of potential problems and management of patient and family preferences.

Care and Treatment Plans, Guidelines and Protocols maintenance and display of guidelines in a general and patient specific context. Also, support for patients enrolled in research protocols, consistent population health-care and patient self-care.

Medication and Immunisation Management support for drug interaction checking, dosing and warnings, medication recommendations and alerts to support safe administration.

Orders, Referrals, Results and Care Management support for management of customisable order set templates and support for result interpretation. Also support for referrals processing and care delivery.

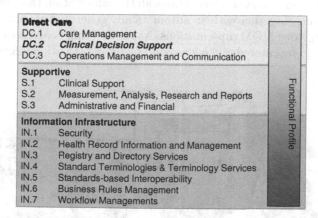

Fig. 4. HL7 EHR-S Functional Model and Functional Profile specification scheme. This Figure lists the first high level set of functions and illustrates how the functional profile consists of a model cross-section, selecting those functions that apply in each case.

Preventive Care and Wellness in the form of alerts (during encounters) and notifications (between encounters).

Population Health support for epidemiological investigations, alerts and notifications for specific at risk population.

Knowledge Access evidence-based knowledge at the point of care for use in health-care decisions and planning, and also access to knowledge by the patient.

In its basic application the EHR-S FM is used to specify a system functional profile (roughly a cross-section of the model, see Figure 4) of an application interfacing with the EHR-S. We employ the EHR-S FM to acquire the necessary framework to drive the requirements on the system and interoperability for information portability between the DSS and the EHR-S.

3   Results and Analysis

3.1   Breast Cancer MDM and MATE

MATE performs diverse tasks, from data input to statistical prognostication and logical decision support. In each MDM patient presentation, the user experience gravitates around the patient for whom investigations are collected and recommendations are suggested (Figure 5). The clinical staff confirms those recommendations or could choose otherwise, as in the case of patient preferences against the recommended treatment. Currently the system is being routinely used in the Breast Cancer MDM at Royal Free Hospital, London UK, in the context of a clinical trial. The trial aims to ascertain how compliance with clinical guidelines by the MDM staff is affected by the use of the tool.

During MDM the summary screen (Figure 6) serves as the basis for the case presentation. The summary screen contains all the information that was available before the meeting, together with an automatically generated graphic sketch and references to previous MDM presentations. As the meeting progresses, additional information could be entered and corresponding prognostication information can

Fig. 5. MATE user experience

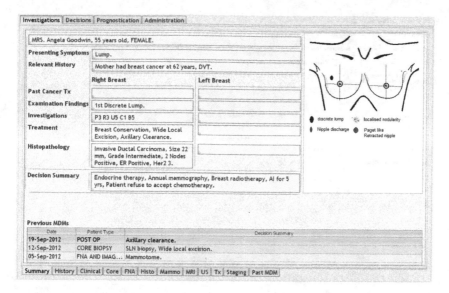

Fig. 6. MATE case summary screen

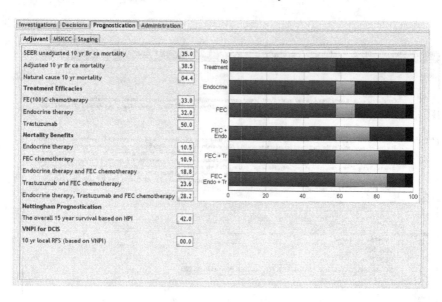

Fig. 7. MATE example prognostication screen

be consulted, in this case Figure 7 demonstrates Adjuvant! Online style prognostications [7]. As shown in Figure 8 recommended and not recommended candidate decisions are brought to the attention of the MDM by highlighting them in blue and red respectively. Inapplicable candidate decisions are grayed out in the user interface. Nevertheless, at any time the user can override the system's

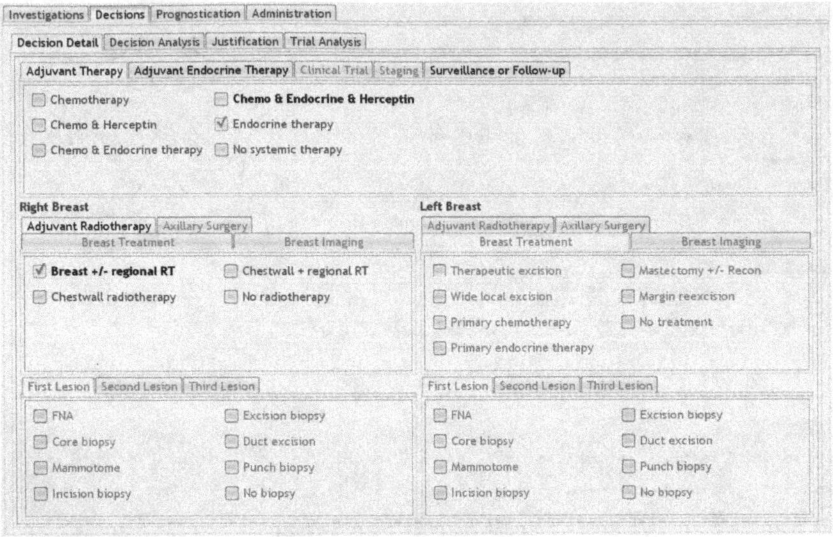

Fig. 8. MATE decisions screen

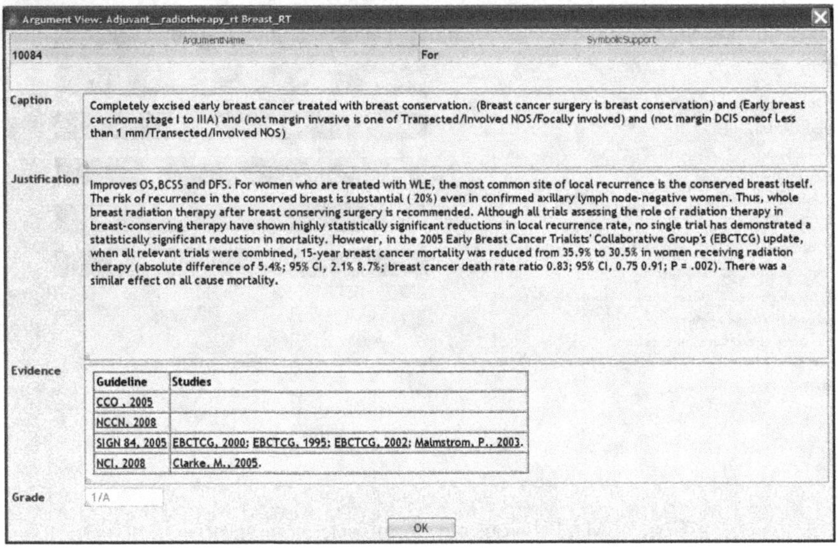

Fig. 9. MATE argumentation screen

recommendation. Finally, the underlying rationale for the recommendation, i.e. the argumentation and underpinning evidence, is accessible to the user for each candidate decision in a separate screen, where arguments for and against are made explicit and the studies are linked via web links as shown in Figure 9.

3.2   Decision Support and Interoperability

The view HL7 EHR-S FM takes on the use of guidelines is a static one. It describes a set of functional requirements for decision support, but none of these are driven by the guidelines themselves. Although this might seem natural from a system specification perspective, it nevertheless contrasts with the view of the computerised clinical guideline community, in the sense that decision support functions, and many others, are outlined and specified by clinical guidelines. The EHR-S FM specifies context-sensitive display of guidelines, but decision support prompts, alerts and notifications bear no explicit relationship to them. We, on the other hand, take the view of an EHR-S FM where the CDS functions are specified by clinical guidelines.

What is important about EHR-S FM is that it allows us to anticipate the functionalities that MATE should possess in order to interoperate with an EHR-S. Consequently, we concentrated on the direct care functions pertaining to CDS and proceeded to group them into logical units as follows:

Authentication & Security functions to ensure secure interaction with the EHR-S, providing authentication and access control.

Messaging HL7 messaging service between the DSS and the EHR-S.

Persistence secure data retrieval and storage for structured and unstructured information and managing data synchronisation between the DSS and the EHR-S.

Translation controlled vocabulary usage and terminology services to facilitate semantic interoperability, providing also vocabulary maintenance and versioning.

Maintenance focused on creation, edition, update and deletion of guidelines.

Auditing retrospective application and analysis of DSS recommendations and providing records to ensure non-repudiation of interactions and attestation of information.

Interchange functions to facilitate importing/exporting of information in standard formats, providing maintenance of the formats and control over interchange agreements.

These group labels reflect our understanding of the requirements for a service oriented architecture where each component corresponds to one of the above functional groups, as shown in Figure 10. Currently, *Tallis* architecture allows the implementation each of these services as a plug-in which can respond to events generated by the *Tallis* guideline enactment engine. Such a loosely-coupled architecture offers modularity and extensibility with clear interfaces. This approach contrasts with that followed by other systems, which use a tightly-coupled design

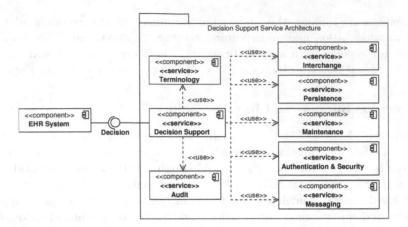

Fig. 10. Proposed service based architecture UML component diagram for a DSS based on the HL7 EHR-S FM. The Figure depicts a set of independent components that allow the decision support service to interact through the **Decision** interface with the EHR-S.

to implement such services. We will discuss this further in Section 3.3. Although an effective approach, it is our view that it makes the process of adding additional services by third-parties cumbersome.

3.3 Related Work

The closest research initiative found in the literature is OncoDoc2 DSS for breast cancer MDM developed by Seroussi et al. [14,15]. The system provides decision support in the form of supervised navigation of a decision tree. Although the system is used in a real hospital setting, it does not considers interfacing with the hospital EHR-S. Guideline compliance is also conceptually different from MATE's view, since in order to arrive to a justified recommendation, the user has to adjust patient's investigations. Most noticeable, the system makes no use of a controlled terminology.

Maviglia et al. [16], depart from a service-based architecture and instead extends GLIF [17] to make the guideline executable within the EHR-S. In summary, the GLIF decision object was extended to provide data binding to the EHR, the action object was extended to provide hooks to various reporting and ordering programs as well as to allow time and event dependent actions. Interestingly, an eligibility action was added to specify if a guideline was appropriate for a particular patient. Also a messaging notification and a questionnaire object were implemented to allow the clinician to provide data not available in the EHR. All this results in a guideline-based, tightly-coupled DSS system acting from within the EHR-S.

Also implemented using GLIF is the work by Laleci et al. [18], but the system architecture is altogether different. It exhibits a service-oriented architecture, by extending GLIF with an EHREntity object which contains attributes enabling it to locate and access the data within the EHR-S. In addition, a terminology server is implemented, not only for data items but also to enhance communication with clinical workflows by defining a functionality ontology and extending GLIF accordingly. This ontology is used to annotate the operational semantics of the clinical services and an ontology mediation service is used during deployment, allowing service discovery even in the presence of different operational semantics.

Hrabak et al. [19] used the SAGE workbench [20] to develop interoperable clinical guidelines for immunisation decision support in primary care. The experience highlights the tremendous effort, common to all initiatives, that controlled terminology usage involves, which is fundamental for interoperability.

An integrated architecture is proposed by Kazemzadeh et al. [21], whereby decision support comes from a blend between guidelines and mined knowledge. The knowledge mining happens offline at different sites, so an intermediate standard interchange format was used (PMML). Together the knowledge and the HL7 Clinical Document Architecture (CDA) are used as data items to the GLIF-based guideline for making recommendations to clinical staff.

Johnson et al. [22] investigates the Virtual Medical Record as a mediation service model between the DSS and the HL7 RIM. In their work they assume that guidelines can be encoded using an uniform virtual electronic health record in conjunction with a terminology service to ensure semantic interoperability. They analyse how the HL7 RIM can be used as a basis for a standardised health record. Their approach to interoperability is data-driven in contrast to ours which is function-driven.

In a departure from all the above initiatives, Schadow et al. [23] propose to use the HL7 Unified Service Action Model (USAM) to represent clinical guidelines in order to enable deployment at the point of care. They argue that, as part of HL7 RIM, USAM provides a sufficiently expressive action-oriented information structure, able to bridge the data mismatch between the DSS and the EHR-S. They further argue that the only solution is to propose a tightly integrated architecture between the EHR-S and the DSS module by means of a shared information model and the HL7 RIM and USAM provides exactly that.

Data mismatch is a well known problem within the DSS community and the solutions are often unsatisfactory, needing additional manual intervention by clinical staff. Proposing a guideline-driven EHR-S data information model is an effective solution to this problem, allowing use of standard interoperability protocols between the EHR-S and the DSS. However, it remains to be seen if imposing minimum additional requirements for data capture by the EHR-S, is desirable or even feasible. Our experience in this regard is extremely positive, as abandoning free text investigations whenever possible, not only allows effective implementation of DSS, but also improves the quality of the information.

4   Conclusions

Cancer MDM together with clinical guidelines and DSS play a prominent role in ensuring high-quality evidence-based health care. However, effective implementation in a real clinical setting could be hindered by lack of interoperability, in particular when interacting with the existing EHR platform. This poses a tremendous technological challenge, requiring the design and implementation of computational frameworks capable of dealing not only with the semantic interoperability, but also with a range of other issues, from knowledge representation to clinical service delivery. This work presented one of such computational frameworks and discussed the methodological considerations to achieve its construction.

The main conclusion of this work is that clinical guideline based DSS can effectively assist breast cancer MDM. An ongoing clinical trial will ascertain how guideline compliance by clinical staff is affected by the use of the system. We have discussed a range of design an implementation issues, fundamental for the deployment in a real clinical setting, from the use of controlled terminology to system architecture and interoperability.

Given that knowledge for breast cancer MDM comes from a compendium of clinical guidelines, we proposed a heuristic to fuse the evidence using the PRO *forma* language and demonstrated how an audit study can be used to ensure correct system recommendations. We addressed interoperability from a functional perspective by proposing a service oriented architecture for decision support based on the HL7 EHR-S FM. From this analysis we argued that the data mismatch problem can be effectively addressed by proposing clinical guideline-based information exchange contracts between the DSS and the EHR-S, where data model contents are dictated by the investigations required for clinical guideline implementation. We are currently investigating how to best specify these contracts based on existing standards.

MATE is routinely used in the context of a clinical trial at Royal Free Hospital, London, UK. The initial results demonstrate that the system is unobtrusive and can cope with the hectic pace of MDMs. We are currently considering the deployment in other hospitals, where a variety of interoperability issues have arisen. In particular institutions which have already invested in MDM information systems, could require only MATE's decision support component, which can be packaged thanks to the loosely-coupled architecture.

Although MATE knowledge base and user interface are specific to breast cancer MDM, the system architecture is sufficiently general for any cancer MDM. We are currently working on a general computational framework for MDM, known as GMATE, that would allow us to roll out this technology to other cancer MDMs and that will facilitate knowledge elicitation and user interface construction.

References

1. Sutton, D.R., Fox, J.: The syntax and semantics of the PROforma guideline modeling language. Journal of the American Medical Informatics Association 10(5), 433–443 (2003)
2. OpenClinical: Methods and tools for the development of computer-interpretable guidelines: PROforma (2006),
   http://www.openclinical.org/gmm_proforma.html (accessed 15 October 2009)

3. Sutton, D.R., Taylor, P., Earle, K.: Evaluation of PROforma as a language for implementing medical guidelines in a practical context. BMC Medical Informatics and Decision Making 6(20) (April 2006), doi:10.1186/1472-6947-6-20

4. Patkar, V., et al.: Breast cancer referral guidelines (2008), http://www.cossac.org/projects/credo/applications#guidelines (accessed 15 October 2009)

5. Golbeck, J., Fragoso, G., Hartel, F., et al.: The national cancer institutes thésaurus and ontology. Web Semantics: Science, Services and Agents on the World Wide Web 1(1), 75–80 (2003)

6. Guyatt, G., Gutterman, D., Baumann, M.H., et al.: Grading strength of recommendations and quality of evidence in clinical guidelines. Chest 129(1), 174–181 (2006)

7. Ravdin, P.M., Siminoff, L.A., Davis, G.J., et al.: Computer program to assist in making decisions about adjuvant therapy for women with early breast cancer. Journal of Clinical Oncology 19(4), 980–991 (2001)

8. Van Zee, K.J., Manasseh, D.M., Bevilacqua, J.L., et al.: A nomogram for predicting the likelihood of additional nodal metastases in breast cancer patients with a positive sentinel node biopsy. Annals of surgical oncology 10(10), 1140–1151 (2003)

9. Patkar, V., Fox, J.: Clinical guidelines and care pathways: A case study applying proforma decision support technology to the breast cancer care pathway. In: Ten Teije, A., Miksch, S., Lucas, P. (eds.) Computer-based Medical Guidelines and Protocols: A Primer and Current Trends. Studies in Health Technology and Informatics, vol. 139, pp. 233–242. IOS Press, Amsterdam, doi:10.3233/978-1-58603-873-1-233

10. Groot, P., Hommersom, A., Lucas, P., et al.: Using model checking for critiquing based on clinical guidelines. Artificial Intelligence in Medicine 46, 19–36 (2009)

11. Perez, B., Porres, I.: Verification of clinical guidelines by model checking. In: 21st IEEE International Symposium on Computer-Based Medical Systems, pp. 114–119. IEEE CS Press, Los Alamitos (2008)

12. Reenskaug, T.M.H.: The original MVC reports. Technical report, Xerox PARC (December 1979), http://heim.ifi.uio.no/~trygver/2007/MVC_Originals.pdf (accessed 15 October 2009)

13. HL7: Electronic health record functional model (2008), http://www.hl7.org/EHR/ (accessed 15 October 2009)

14. Séroussi, B., Bouaud, J., Gligorov, J., Uzan, S.: Supporting multidisciplinary staff meetings for guideline-based breast cancer management: a study with oncodoc2. In: AMIA Annual Symposium Proceedings, pp. 656–660 (2007)

15. Séroussi, B., Bouaud, J., Antoine, E.C.: Oncodoc: a successful experiment of computer-supported guideline development and implementation in the treatment of breast cancer. Artificial Intelligence in Medicine 22, 43–64 (2001)

16. Maviglia, S.M., Zielstorff, R.D., Paterno, M., et al.: Automating complex guidelines for chronic disease: Lessons learned. Journal of the American Medical Informatics Association 10(2), 154–165 (2003)

17. Boxwala, A.A., Peleg, M., Tu, S., et al.: Glif3: a representation format for sharable computer-interpretable clinical practice guidelines. Journal of Biomedical Informatics 37, 147–161 (2004)

18. Laleci, G.B., Dogac, A.: A semantically enriched clinical guideline model enabling deployment in heterogeneous healthcare environments. IEEE Transactions on Information Technology in Biomedicine 13(2), 263–273 (2009)

19. Hrabak, K.M., Campbell, J.R., Tu, S.W., et al.: Creating interoperable guidelines: Requirements of vocabulary standards in immunization decision support. In: Klaus, A., Kuhn, J.R., Warren, T.Y.L. (eds.) MEDINFO 2007 - Proceedings of the 12th World Congress on Health (Medical) Informatics Building Sustainable Health Systems. Studies in Health Technology and Informatics, vol. 129, pp. 930–934 (2007)
20. Tu, S.W., Campbell, J.R., Glasgow, J., et al.: The sage guideline model: Achievements and overview. Journal of the American Medical Informatics Association 14(5), 589–598 (2007)
21. Kazemzadeh, R.S., Sartipi, K.: Interoperability of data and knowledge in distributed health care systems. In: Proceedings of the 13th IEEE International Workshop on Software Technology and Engineering Practice, pp. 230–240. IEEE Computer Society, Los Alamitos (2005)
22. Johnson, P.D., Tu, S.W., Musen, M.A., Purves, I.: A virtual medical record for guideline-based decision support. In: AMIA Annual Symposium Proceedings, pp. 294–298 (2001)
23. Schadow, G., Russler, D.C., Mead, C.N., McDonald, C.J.: Integrating medical information and knowledge in the hl7 rim. In: AMIA Annual Symposium Proceedings, pp. 764–768 (2000)

Technical Solutions for Integrating Clinical Practice Guidelines with Electronic Patient Records

Silvia Panzarasa[1], Silvana Quaglini[2], Anna Cavallini[3], Giuseppe Micieli[3],
Simona Marcheselli[4], and Mario Stefanelli[2]

[1] CBIM Consorzio di Bioingegneria e Informatica Medica, Pavia, Italy
[2] Department of Computer Engineering and Systems Science, University of Pavia, Italy
[3] IRCCS C. Mondino Foundation, Pavia, Italy
[4] IRCCS Humanitas, Rozzano (MI), Italy
silvana.quaglini@unipv.it

Abstract. The success of a decision support system based on clinical practice guidelines does not only depend on the quality of the decision model used to represent and execute guideline recommendations, but also on the design of interactions of the system with the end-user interface and the electronic patient record. This paper describes technical solutions adopted to add decision support functionalities to two existing information systems for stroke patients. Despite the specific medical application, the approach is quite general, relying on two main functionalities: a real-time decision support system based on workflow technology (careflow) and an off-line tool for non-compliance detection, called "Reasoning on Medical Action" (RoMA). The integration has been developed maintaining independence between data management and knowledge management, and minimizing changes to existing user's interfaces. The paper illustrates in particular the middleware layer created to allow communication between the evidence-based system and the electronic patient record.

Keywords: workflow management systems, clinical practice guidelines, middleware, electronic patient record.

1 Introduction

Paper-based clinical practice guidelines (GLs) are difficult to be adequately distributed and used by end users [1,2]. Medical informatics community hypothesized that more *formal* electronic versions would increase both diffusion and physicians' compliance with GLs. As a matter of fact, formal models, such as flowcharts or other graphical metaphors, are likely to provide a more friendly and immediate way of reading a GL, thus improving learning and internalization of its content. These models can be profitably exploited also within educational/simulation tools [3]. Moreover, formal models allow easier GL integration with the electronic patient record (EPR), raising possibility of building real-time decision support systems (DSS), that should further improve physicians' compliance. Unfortunately, medical informatics solutions did not always succeed, thus raising doubts about the rationale they were based on. Concerning the compliance improvement, some favorable evidence exists in the

D. Riaño et al. (Eds.): KR4HC 2009, LNAI 5943, pp. 141–154, 2010.
© Springer-Verlag Berlin Heidelberg 2010

literature [4, 5, 6], but also contradictory results are reported [7, 8, 9]. Usability issues are often a confounding factor in studies attempting to test the effectiveness of these systems. Therefore, while in past years the emphasis was on how to formally represent a GL [10], more recently it is on how to best integrate GLs in real healthcare settings, taking into consideration the local work organization and the different needs and constraints that users perceive in their daily practice [11]. We believe that a DSS must be integrated into the EPR to be really useful: if not, problems of multiple data input, and users training to different systems are very likely to impair its acceptability. Under this assumption, the following, albeit non-exhaustive, list of questions gives an idea of the number and the variety of issues to be considered before implementing the system:

GL navigation opportunities - While using the EPR interface, should the user be provided with facilities to navigate a formal representation of the GL? If not, the GL engine will *simply* run behind the system and the user will only see the resulting recommendations. If navigation is provided, (a) should a GL be shown wholly or partially, e.g. only the portion related to the actual patient's condition? (b) when does navigation apply? On user's request or automatically, according to some trigger?

User notification - How to show the GL recommendations? By a message in a pop-up window (synchronously)? Or by a less intrusive communication to be read (asynchronously) at the user's willingness? Should alternative communication modalities be considered, e.g. SMS, in order to reach users even when they are not in front of a workstation? How to manage notifications to multiple users?

Non-compliance management- If a non compliance is detected, does the user have to provide a motivation? If yes, when? Immediately or when he decides to do it, or at a pre-defined event, for example daily or weekly meeting (if any) or at the patient's discharge? Who must be notified about non-compliance?

Answering these questions is particularly challenging when an information system already exists, and users are familiar with its interface. In this case it is very inconvenient for users to shift to other systems or to use parallel ones. Medical informatics and artificial intelligence experienced several failures in the past, due to this problem. Learning new interfaces, multiple data input, boring interaction modalities have been the main causes of early abandon of several information systems and expert systems in medicine [12]. To overcome these problems, while developing the application described in this work, the driving idea has been that users must perceive the new system as an upgrade of the previous one, just providing some additional functionalities to the usual interface. All steps of the new system design and development have been performed in collaboration with the technical team responsible for the EPR and the healthcare personnel of the involved medical units. We stress the paramount importance of knowledge and skill sharing among working teams, to give the final tool a high probability of success and users' satisfaction.

The proposed methodologies and technologies are illustrated in the area of stroke management. The generality of the approach is demonstrated by integrating the same DSS with two different EPR schemas in two Stroke Units (SUs) in Lombardy, namely Istituto Clinico Humanitas ICH in Rozzano and Hospital Mondino in Pavia. This is a *technical* paper. Section 2 illustrates the project background, and the two EPRs that have been integrated with the DSS; Section 3 shows how GL recommendations are formalized into computational rules and the results of the DSS are obtained; Section 4 illustrates the integration with the EPRs; Section 5 illustrates the user interface; Section 6 compares our project with related works and Section 7 outlines conclusions and future work. Answers to the above listed questions are provided in Sections 3 and 5.

2 Background

The STAGE (Stroke Active Guideline Evaluation) project was funded some years ago by the Italian Ministry of Health, with the aim of improving the diffusion of best practices in stroke management. The project promised to provide about 20 neurological departments with a DSS, based on "The Italian SPREAD guidelines for stroke prevention and management" [13], delivered by the Italian Stroke Forum (www.strokeforum.org). Some hospitals adopted an EPR developed by the ICH Information System Department, others were equipped with an EPR implemented with the commercial tool WINCARE® (by TSD Projects). The two systems had very similar graphical user interface (GUI), based on forms for data collection and textual reports generation. Usernames and passwords were grouped on the basis of the role (physician, professional nurse, physiotherapist, etc), allowing the visualization of competent forms only. Fig. 1 illustrates the GUI without any DSS integration. The list of forms (bottom left) is simply ordered by name (as in the figure), or by date, and there is no *intelligent* tailoring to the specific patient.

Despite a similar interface, the two EPRs have a different structure (see portions of Entity-Relationship diagrams (ERDs) in Fig. 2) and implementation modality:

• Humanitas EPR is a web-based system, relying on Oracle, in which all the data, coming from different forms, are stored in a unique table containing: a unique id (primary key), a field id, a form id (foreign keys for tables describing each data item

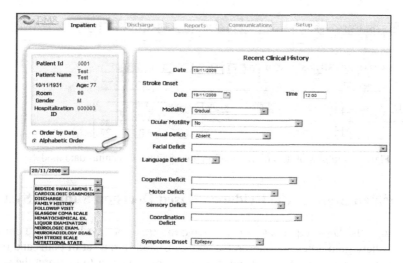

Fig. 1. The graphical user interface of the EPR in use at Istituto Clinico Humanitas-Stroke Unit, before decision support was integrated

and the form to which it belongs), a value field, date of value input, patient id and hospitalization id. A data item is defined by an identifier, its description (that is also the label in the GUI), and type (number, text, encoded, etc.). Fig. 3 shows how different data types (an encoded item and a numerical one) are stored;

• Wincare® is a stand-alone application built on a MS Windows platform that stores information in a "classical" relational DB structure, in which each table represents a form of the clinical chart (i.e. one table for Personal Data, one for Past History, one for Laboratory data, etc). Each record in a table is identified by patient id, hospitalization id, date of execution (primary key) and as many additional attributes as the data entry fields in the corresponding form. The DB still relies on Oracle.

Both systems use standard codes for pathologies (ICD9-CM) and drugs (active principle unique identifier) and restrict free text input to very few items.

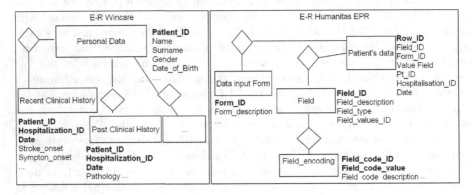

Fig. 2. Portions of the ERDs of the two database implemented in the different hospitals

Data input Form

FORM_ID	FORM_DESCRIPTION
39	Cardiological Examination
23	Objective Examination

Field_encoding

FIELD_CODE_ID	FIELD_CODE_VALUE	FIELD_CODE_DESCRIPTION
DC_ECG	2	Atrial Fibrillation

Field

FIELD_ID	FIELD_DESCRIPTION	FIELD_TYPE	FIELD_VALUES_ID	FORM_ID
676	referral_ECG	pop_up_menu	DC_ECG	39
345	Weight (kg)	number		23

Patient's data

ROW_ID	FIELD_ID	FORM_ID	VALUE_FIELD	PT_ID	HOSP_ID	DATE
27250	676	39	2	355	781	12-11-2009 11:00
27320	345	23	83	355	781	12-11-2009 10:20
27325	345	23	80	355	781	19:11:2009 11:40

Fig. 3. Examples of data representation through the Humanitas data model

3   Implementation of the Evidence-Based Decision Support System

Implementing the DSS requires two main steps: the first one is a knowledge conversion step, where guidelines are interpreted and translated into a set of paths and rules. Then, one or more inference engines must be developed in order to interpret the above formalization and build the computational tools.

3.1 From GL Recommendations to Production Rules

The information, expressed in natural language in the GL text, has been converted into IF-THEN rules. This process is very critical and requires strong collaboration between physicians and knowledge engineers. As a matter of fact, the GL text often is ambiguous and includes some "hidden" knowledge, that is not explicitly written because it is part of the cultural background of any physician (e.g. which tests belong to a given diagnostic work-up, which diseases belong to a certain category, etc) or simply because obvious (e.g. the patient must be alive at a certain time for being evaluated). As an example, let us consider Recommendation 5.5 from the diagnostic work-up chapter of the SPREAD Guideline:

5.5 After a TIA or a stroke, transthoracic echocardiography (TTE) is recommended only when a heart disease is clinically suspected

In order to build the rule, we must associate a sensible temporal frame to the word "After" and we must know which are the patients' data suggestive of "heart disease". Once this information has been elicited from physicians, a semi-formal version of the rule is obtained:

"IF the patient is not dead within 3 days from the admission **AND** the patient shows one out of (myocardial infarction, atrial fibrillation, coronaropathy, heart failure, arrhythmias, prosthesis) **THEN** TTE must be done"

According to the common physicians' decision process and used terminology, a rule antecedent is in general composed by "exclusion criteria" (e.g. the fact that the patient must be alive at three days in the above rule) and "inclusion criteria" (e.g. presence of myocardial infarction). To represent them, it is very useful to define reusable data abstractions, to be exploited within several rules. For example, "Early_death" (death within 3 days from admission) has been defined as an abstraction that combines the hospital stay with the discharge status, represented by the Rankin scale (ranging from 0=well to 6=death) . Once matched with the data available in the EPR and the defined abstractions, the rule eventually is:

```
IF NOT (early_death)
AND    (myocardial_infarction  OR  atrial_fibrillation  OR
coronaropathy OR heart_failure OR arrhythmias)
THEN   TTE_execution
```

The same process has been done for all the recommendations, obtaining both the list of those that can be interpretable (converted to IF-THEN rules) and the list of data items that EPR and DSS must share (the "minimum data set"). This process can lead to update the EPR data model, with the aim of increasing the number of recommendations that can be interpreted. The computational representation of a rule exploits relational tables, as shown in Fig. 4 for the above rule.

The value of a CRITERIUM in records with TYPE_CODE=2 (raw data) of *Exclusion_Criteria, Inclusion_Criteria* and *Action_Criteria* tables, and the value of the DATA_ITEM attribute in *Abstraction_Data* is the "standard" name given to the medical concept. To be precise, it is a SNOMED code, but for sake of clarity we report its description, e.g. "Atrial fibrillation (disorder)" should be the SNOMED code D3-31520. When no unique code has been found for a composite concept (e.g.

Rankin at discharge, which includes a scale assessment and its timing), we used a combination of codes: the assessment code and the point-in-time code as shown in *Abstraction_Data*. In this way it is possible to distinguish the pre-stroke Rankin scale from the same scale measured at emergency department and at discharge. *Abstraction_Rule* contains, for each CRITERIUM, the corresponding RULE defined using the standard medical concept (DATA_ITEM).

Data defines the table and the attribute where the actual value of DATA_ITEM is stored. The DSS uses these tables to activate the recommendation actions when no exclusion criteria is present and at least one inclusion criteria is present, considering that a criterion can also be a logical product of conditions (not shown in the above tables). We remark that the DSS does not access directly the EPR, gathering the patients' data from a middleware layer, that will be illustrated in Section 4.

Exclusion_Criteria

RECOMM.	GL	CRITERIUM	TEXT		TYPE_CODE
5.5	SPREAD	Early_death	Patient is dead within 3 days		1

The values of TYPE_CODE are defined in Types.

Inclusion_Criteria

RECOMM.	GL	CRITERIUM	TEXT	TYPE_CODE
5.5	SPREAD	Myocardial infarction (disorder)	Patient with MI	2
5.5	SPREAD	Atrial fibrillation (disorder)	Patient with Atrial fibrillation	2
5.5	SPREAD	Coronary arteriosclerosis (disorder)	Patient with coronaropathy	2
5.5	SPREAD	Heart failure (disorder)	Patient with Heart failure	2
5.5	SPREAD	Ventricular arrhythmia (disorder)	Patient with ventricular arrhythmia	2

Action_Criteria

RECOMM.	GL	CRITERIUM	TEXT	TYPE_CODE
5.5	SPREAD	Transthoracic echocardiography (procedure)	Trans-Thoracic echocardiography must be done	2

Types

CODE.	DESCRIPTION
1	Abstraction
2	Raw data

Abstraction_Rule

CRITERIUM.	RULE
Early_death	Rankin activities of daily living scale (assessment scale) @ Patient discharge (procedure)=6 AND (Date (attribute) @ Patient discharge (procedure) - Date of admission (observable entity)<=3

Abstraction_Data

CRITERIUM.	DATA_ITEM
Early_death	Date of admission (observable entity)
Early_death	Date (attribute) @ Patient discharge (procedure)
Early_death	Rankin activities of daily living scale (assessment scale) @ Patient discharge (procedure)

Data

DATA_ITEM.	TABLE	FIELD
Date of admission (observable entity)	Admission	Admission_Date
Date (attribute) @ Patient discharge (procedure)	Discharge	Discharge_Date
Myocardial infarction (disorder)	Past clinical history	Myocardial_infarction
Atrial fibrillation (disorder)	Past clinical history	Atrial_fibrillation

Fig. 4. The representation of the recommendation 5.5 in the DSS database

3.2 The Inference Engines

As already mentioned, our solution is based on two components: a real-time DSS and a tool running at patient's discharge to detect non-compliance with GL.

The real-time system, called "Careflow Management System (CfMS)" [14,15] is built on top of Oracle Workflow™. The Workflow logic, i.e. the possible clinical paths, is grounded on the SPREAD GL. Fig. 5 shows two portions of the careflow model. The CfMS engine elaborates patient's data according to GL rules, suggests the actions to be done (the so-called *to-do-list*, Fig. 5a) and communicates the patient-specific recommendations to the users (see *notification*, Fig. 5b). In order to manage these notifications, the fundamental knowledge of organisational aspects of the health structure must be modeled: roles, skills and both human and instrumental resources. This is done using a set of relational tables storing, for every resource, its type, location, availability during the 24 hours, associated roles, etc.

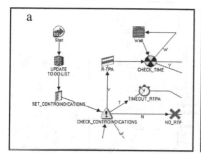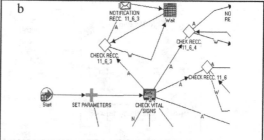

Fig. 5. Two portions of the careflow model: (a) the process for r-tPA treatment with contraindication checking and (b) complications management: as soon as the vital signs are collected, either manually or with monitor instrumentation, depending on their values, different recommendations are checked

Suggestions and recommendations shown to users in real time can be accepted or not, but no motivations or comments are requested from users during the clinical routine. The off-line reasoning system, called RoMA (Reasoning on Medical Actions), runs at discharge, when all patient data have been entered in the EPR and physicians have time to summarize and reason about the whole care process.

For each patient, a set of queries, based on the rules described in Section 3.1, is activated and RoMA creates a report with (a) missing data, with respect to the minimum data set required for GL interpretation, as defined in section 3.1; (b) the list of recommendations the patients was eligible for; (c) the list of non-compliances. In case of missing data or non-compliance, users may provide motivations, selecting from a predefined taxonomy, and comments in free text [16]. The taxonomy has been developed according to the feedbacks that we expect to obtain and report to responsible persons or units, hopefully able to manage the problem. Thus the user may highlight organization-related problems, producing feedback for healthcare administrators; medical-type problems, producing feedback for the personnel involved in the care process and guideline reviewers; and technical problems, producing feedback for information systems managers. Other possible causes of non-compliance may be highlighted, such as GL formalisation problems, i.e. the user claims that the computerised recommendation is not a correct interpretation of the GL text. This last problem is likely to appear during the first implementation phase, and the feedback is useful for revisiting the GL text with medical experts, to fix possible misinterpretations.

4   Integration with the EPR

To keep the systems (Evidence-based DSS and EPR Management System) independent while granting communication, a middleware layer (see Fig. 6) has been developed. It is composed by a supporting database (Oracle DB) and an Interpreter written in PL/SQL.

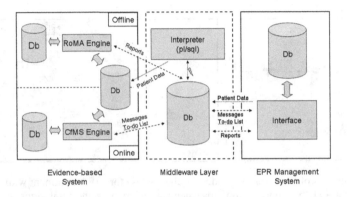

Fig. 6. Architecture of the integrated system

The middleware DB (see E-R diagram in Fig. 7) includes a set of tables with static content (grey in the figure) and a set of tables with patient- and time-specific content. A key attribute is Task_ID, matching the CfMS "task" concept with the corresponding EPR form (EPR_form_name). In fact, executing a task implies filling a data form, whose compilation will then be used to check for that task's execution.

4.1   Patient's Data Flow and Message Flow

As shown in Fig.7, once a patient is admitted to a Stroke Unit, through the Admission form, a new instance of the CfMS is activated by the interpreter and the corresponding patient ID is inserted into *Active-Patients*. Then, whenever new data enter the EPR for an "active" patient, the attributes that are useful for the GL interpretation are transferred into *Data-exchange*, according to the information stored in *Transcoding-Tasks*. It contains, for each EPR form, the names of the required attributes, as they appear in the EPR itself, separated by the special character "§", e.g. weight§heigth§systolicBP§dyastolicBP for Wincare and 345§346§348§351 for Humanitas. The content of this table is different for every installed Middleware, since it refers to the specific EPR data model and attribute names.

We remark that every future change of the required attributes or forms, for example due to periodic updates of the SPREAD GL, will only imply an update of the static tables, for example adding new attributes to an existing or new form.

As a new record is inserted into *Data-exchange* (e.g. "80§175§140§95" in the form "Objective Examination" for the patient "P1" on "December 12th 2008"), a trigger activates the Interpreter in order to create a dynamic SQL query storing patient data in the correct table of the DSS database. This can be done using *Mapping*, which contains the name of this table, e.g. "OBJ_EXAM" and the corresponding sequence of attributes (e.g. "P_WEIGHT§ P_HEIGHT§P_SYSBP§P_DYABP"). According to the previous example, the query generated by the Interpreter will be "insert into OBJ_EXAM (P_ID, DATE, P_WEIGHT, P_HEIGHT, P_SYSBP, P_DYABP) VALUES ('P1', '12/12/2008', '80', '175', '140', '95'). In the same way, any update of an EPR form triggers the Interpreter to run an Update SQL query to synchronize the DSS patient database, which is accessible by both CfMS and RoMA engines.

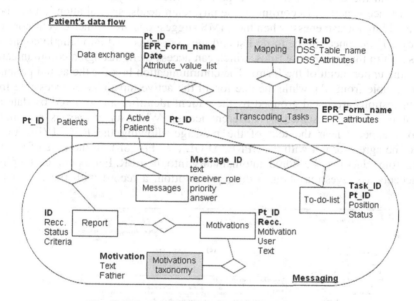

Fig. 7. E-R diagram of the Middleware Database

When a discharge form is filled out for an active patient, he/she is deleted from *Active-Patients* by the Interpreter and the related CfMS instance is deactivated.

As shown in Fig. 6, the patient's data flow is unidirectional, from EPR to DSS.

On the other side, the DSS needs to communicate with the EPR in order to exploit its interface for showing the results of data processing, both in real time (by the CfMS) and offline (by the RoMA module). When the CfMS generates a suggestion, or a new to-do list, tables *Messages* and *To-do-list* are filled. To refresh the user's interface, the EPR executes a reading of these tables periodically (every minute in our implementations). Opposite to the patient's data flow, the message flow is bi-directional. As a matter of fact, the CfMS requires that users answer its communications. These answers are stored in *Messages* by the EPR. A user may also modify the to-do list by adding forms different from those proposed by the CfMS, thus affecting the table *To-do-list*.

The RoMA module, as already said, is activated at patient discharge by passing the patient identification. After the elaboration, a report is produced (stored in *Report* table) with the list of recommendations the patients was eligible for and the non-compliances detected. The EPR reads the *Report* table and shows the output, prompting the users for providing motivations (*Motivations* table) choosing them from a predefined taxonomy (*Motivation_Taxonomy* table).

5 The Smart User Interface

Fig. 8 illustrates how the form in Fig. 1 has been turned into a *smart form*. Part (A) is the area with the details of the actual patient; when it is yellow it indicates that a CfMS instance is active. According to the physician needs, special attention has been put in avoiding intrusiveness: when the CfMS suggestions are neither urgent nor critical, they do not appear in the to-do list and they are translated into simple communications sent to the appropriate roles, which can access them through a communication box (in the upper menu of the form). The communication box for the actual patient is also accessible from (B), while the one for all the active inpatients is accessible from (C). Suggestions are listed grouped by the patient identifier and ordered by date and level of priority (represented by different icons). When an answer is required, a checkbox appears near the text of the message allowing the RECEIVER to say whether he agrees or not with the SPREAD GL (see Fig. 8F). In the case of positive answer, some data may be stored automatically into the EPR. For example, if a physician accepts a suggestion about a drug prescription, a record with the drug name,

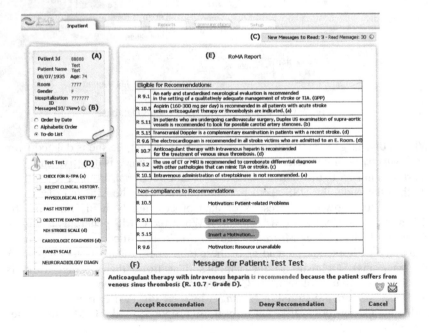

Fig. 8. The new interface of the EPR integrated with the EB system

dosage, and timing is stored and at the same time a message is sent to the role "nurse", which is responsible for drug administration.

Fig 8D is the to-do list as it comes from the CfMS: icons of different colour (red, green and yellow) represent the different status of the tasks (*exception, executed, under execution*), while absence of an icon means that the task has still to be done. Once the corresponding form is accessed, the fields useful for GL interpretation are highlighted to foster their compilation. In order to avoid over-information in urgent situations, only the forms useful to tackle the urgency are shown. Thus, with respect to Fig. 1, where the list of forms was static and unspecific, now it is patient-tailored and dynamic, because it is built on the patient's data and CfMS suggestions.

The RoMA report is visualized in Fig.8E with the list of the recommendations the patient was eligible for and the non compliances, in this example partially filled with the corresponding motivation. Together with the STATUS, the grade of recommendation is also shown, as it appears in the SPREAD GL, by letters a-d.

By clicking the recommendation number, the user can access directly the online version of the SPREAD GL, at the page related to that recommendation, and can exploit the navigation facilities provided by the SPREAD website.

6 Comparison with Related Works

Several issues described in this paper have been considered by other researchers. One of them is GL text interpretation. Reading a GL, we realise that there is so much missing or implicit knowledge and even inconsistencies (either real or apparent, due to text ambiguity) that it is impossible to formalise it without the collaboration of medical experts. Moreover, often, as well described in [17], interpretation by multiple medical experts is different from each other. A method for eliciting medical experts' knowledge can be found in [18], where a tool that support the process of GL formalization is described, based on the *many-headed bridge* idea and on an intermediate representation of GL dimensions, such as control and data flow, scientific evidence, temporal aspects, resources and patient conditions. Another GL representation model is proposed in [19] to deal with ambiguity in GL text and with the process of acquiring the knowledge missing in the text but required for the GL implementation. In our approach, the control and data flow is represented through the workflow model and recommendations are specified as IF-THEN rules. We pushed physicians to interpret the GL text and describe the rule condition as a product of inclusion and exclusion criteria, since this is their common reasoning schema. These criteria, still expressed in natural language, are then translated into logical combinations of conditions on the EPR attributes. Then, during the very early implementation phase, RoMA is used as a validation tool for the implemented rules: a user could be non-compliant with a system-generated suggestion because the implemented rule is not correct (or at least it seems not correct to him); in this case, the user motivates his non-compliance by choosing "Guideline formalization problem" in the non-compliance taxonomy. A motivation of this type is notified to the GL implementation responsible, that will take care of the rule revision.

Another issue is the local adaptation of GLs, and their integration with local EPRs: Peleg et al. [20] describe how they supported these processes for a diabetic-foot care

GL, using the GLIF3 modeling language, and matching the obtained formal representation with the EPR schema and local needs. The formalization has then been validated by manual check and execution of test cases. In our project, while the GL recommendations did not require a local adaptation (SPREAD is an Italian collaboration initiative), the careflow model has been tailored to each hospital by means of the organization-related tables mentioned in section. 3.2 and also the middleware layer has been adapted (Transcoding_Tasks in Fig. 7).

Adoption of standard data models (such as HL7-RIM) and standard terminologies (SNOMED, LOINC) to enhance reusing a DSS with different EPRs is an important and critical point, extensively discussed in [21-23]. The idea is that runtime queries formulated using standard terms are automatically translated in queries able to retrieve patient's data in the local DB. The authors also illustrate how mapping GL knowledge to EPR data, and defining data abstractions, can be facilitated by terminology mapping tools and ontologies. Our solution for the data model is close to the Virtual Medical Record proposed in [23], since our middleware contains the minimum data set necessary for GL implementation. Concerning terminology, we used SNOMED to encode those data, even if a problem still to be solved is its Italian translation.

7 Conclusion and Future Work

We described the architecture and the technical solutions used to integrate a DSS with two existing EPR systems. Results observed on physicians' compliance with GLs are behind the scope of this technical paper: in [24] compliance with GLs, before and after the system introduction, has been analyzed, as well as the accuracy of the clinical chart compilation, care process indicators, and system usability, showing that the system enhances the clinical practice without boring users.

A lot of work has still to be done: (a) the relational schema shown in Fig. 4 covered all the recommendations that we implemented for the SPREAD GL, but of course it requires an extensive validation using different GLs; (b) we plan to improve abstractions definition, in particular for representing complex temporal abstractions; (c) the motivations for non-compliance have been used to detect pitfalls in the GL text interpretation, as we mention in Section 3.2.2, but they must be exploited more extensively to provide feedbacks also to non-medical personnel (e.g. to technical staff for instrumentation faults), and automatically (e.g. computing statistics every month and sending reports through the CfMS to the appropriate roles).

References

1. Eccles, M.P., Grimshaw, J.M.: Selecting, presenting and delivering clinical guidelines: are there any "magic bullets"? Med. J. Aust. 180(6 Suppl.), S52–S54 (2004)
2. Prior, M., Guerin, M., Grimmer-Somers, K.: The effectiveness of clinical guideline implementation strategies-a synthesis of systematic review findings. J. Eval. Clin. Pract. 14(5), 888–897 (2008); Review

3. Groot, P., Hommersom, A., Lucas, P.J., Merk, R.J., ten Teije, A., van Harmelen, F., Serban, R.: Using model checking for critiquing based on clinical guidelines. Artif. Intell. Med. 46(1), 19–36 (2009)
4. Kawamoto, K., Houlihan, C.A., Balas, E.A., Lobach, D.F.: Improving clinical practice using clinical decision support systems: a systematic review of trials to identify features critical to success. BMJ. 2, 330 (2005)
5. Moyer, P.: Guidelines for Inpatient Treatment of Stroke Improve Quality of Care. In: 30th International Stroke Conference, New Orleans: Abstract 93 (2005)
6. Wells, S., Furness, S., Rafter, N., Horn, E., Whittaker, R., Stewart, A., Moodabe, K., Roseman, P., Selak, V., Bramley, D., Jackson, R.: Integrated electronic decision support increases cardiovascular disease risk assessment four fold in routine primary care practice. Eur. J. Cardiovasc. Prev. Rehabil. 15(2), 173–178 (2008)
7. Eccles, M., McColl, E., Steen, N., Rousseau, N., Grimshaw, J., Parkin, D., Purves, I.: Effect of computerised evidence based guidelines on management of asthma and angina in adults in primary care: cluster randomised controlled trial. BMJ 325, 941 (2002)
8. Wears, R.L., Berg, M.: Computer Technology and Clinical Work - Still Waiting for Godot. JAMA 293, 1261–1263 (2005)
9. Green, C.J., Fortin, P., Maclure, M., Macgregor, A., Robinson, S.: Information system support as a critical success factor for chronic disease management: Necessary but not sufficient. Int. J. Med. Inform. 75(12), 818–828 (2006)
10. Peleg, M., Tu, S., Bury, J., Ciccarese, P., Fox, J., Greenes, R.A., Hall, R., Johnson, P.D., Jones, N., Kumar, A., Miksch, S., Quaglini, S., Seyfang, A., Shortliffe, E.H., Stefanelli, M.: Comparing computer-interpretable guideline models: a case-study approach. J. Am. Med. Inform. Assoc. 10(1), 52–68 (2003)
11. Middleton, B.: The clinical decision support consortium. Stud. Health Technol. Inform. 150, 26–30 (2009)
12. Coiera, E.: The Guide to Health Informatics, 2nd edn. Arnold, London (2003)
13. SPREAD Stroke Prevention and Educational Awareness Diffusion (2003), http://www.spread.it
14. Panzarasa, S., Bellazzi, R., Larizza, C., Stefanelli, M.: A careflow management system for chronic patients. In: Fieschi, Yu-Chan, C., Li, J. (eds.) MedInfo 2004, pp. 773–777 (2004)
15. Panzarasa, S., Maddè, S., Quaglini, S., Pistarini, C., Stefanelli, M.: Evidence-based careflow management systems: the case of post-stroke rehabilitation. J. Biomed. Inform. 35(2), 123–139 (2002)
16. Panzarasa, S., Quaglini, S., Cavallini, A., Marcheselli, S., Stefanelli, M., Micieli, G.: Computerised Guidelines Implementation: Obtaining Feedback for Revision of Guidelines, Clinical Data Model and Data Flow. In: Bellazzi, R., Abu-Hanna, A., Hunter, J. (eds.) AIME 2007. LNCS (LNAI), vol. 4594, pp. 461–466. Springer, Heidelberg (2007)
17. Shalom, E., Shahar, Y., Lunenfeld, E., Taieb-Maimon, M., Young, O., Goren-Bar, D., et al.: The Importance of Creating an Ontology-Specific Consensus Before a Markup-Based Specification of Clinical Guidelines. In: Proc. ECAI, Riva del Garda, Italy (2006)
18. Seyfang, A., Miksch, S., Marcos, M., Wittenberg, J., Polo-Conde, C., Rosenbrand, K.: Bridging the Gap between Informal and Formal Guideline Representations. In: Proc. ECAI, Riva del Garda, Italy (2006)
19. Shiffman, R.N., Michel, G., Essaihi, A., Thornquist, E.: Bridging the guideline implementation gap: a systematic, document-centered approach to guideline implementation. J. Am. Med. Inform. Assoc. 11(5), 418–426 (2004)

20. Peleg, M., Wang, D., Fodor, A., Keren, S., Karnieli, E.: Lessons learned from adapting a generic narrative diabetic-foot guideline to an institutional decision-support system. Stud. Health Technol. Inform. 139, 243–252 (2008)
21. German, E., Leibowitz, A., Shahar, Y.: An architecture for linking medical decision-support applications to clinical databases and its evaluation. J. Biomed. Inform. 42(2), 203–218 (2009)
22. Peleg, M., Keren, S., Denekamp, Y.: Mapping Computerized Clinical Guidelines to Electronic Medical Records: Knowledge-Data Ontological Mapper (KDOM). J. Biomed. Inform. 41(1), 180–201 (2008)
23. Tu, S.W., Musen, M.A., Shankar, R., Campbell, J., Hrabak, K., McClay, J., et al.: Modeling guidelines for integration into clinical workflow. In: Medinfo 2004, pp. 174–178 (2004)
24. Panzarasa, S., Quaglini, S., Micieli, G., Marcheselli, S., Pessina, M., Pernice, C., Cavallini, A., Stefanelli, M.: Improving compliance to guidelines through workflow technology: implementation and results in a stroke unit. Stud. Health Technol. Inform. 129, 834–839 (2007)

Towards a Possibility-Theoretic Approach to Uncertainty in Medical Data Interpretation for Text Generation

François Portet[1] and Albert Gatt[2]

[1] Laboratoire d'Informatique de Grenoble, Grenoble Institute of Technology, France
francois.portet@imag.fr
[2] Institute of Linguistics, University of Malta, Malta
albert.gatt@um.edu.mt

Abstract. Many real-world applications that reason about events obtained from raw data must deal with the problem of temporal uncertainty, which arises due to error or inaccuracy in data. Uncertainty also compromises reasoning where relationships between events need to be inferred. This paper discusses an approach to dealing with uncertainty in temporal and causal relations using Possibility Theory, focusing on a family of medical decision support systems that aim to generate textual summaries from raw patient data in a Neonatal Intensive Care Unit. We describe a framework to capture temporal uncertainty and to express it in generated texts by mean of linguistic modifiers. These modifiers have been chosen based on a human experiment testing the association between subjective certainty about a proposition and the participants' way of verbalising it.

1 Introduction

Clinical decision support systems (CDSS) run into problems when there is temporal uncertainty or inaccuracy in their input data, which can arise for a variety of reasons. For example, medical staff often record events when they have time to do so, rather than when they actually happened. In addition, existing database management systems tend not to deal with temporal data in a principled fashion [1]. Uncertainty and inaccuracy make the tasks of reasoning about temporal and causal relationships more difficult, especially where input data is provided in a raw form.

Classical CDSS (especially expert systems) typically approach the problem of uncertainty either by restricting output to what the system is completely certain about, or by communicating findings using a ranking mechanism [2]. In contrast to such systems, the family of CDSS being developed in the BabyTalk project [3] aim to provide *textual* summaries of heterogeneous medical data (both automatically and manually entered) to support decisions by carers in Neonatal Intensive Care Units (NICUs). The goal is to communicate the relevant aspects of patient data, by using Natural Language Generation (NLG) techniques to produce a descriptive summary, leaving it up to the user to decide on the best course of action. The emphasis on generating textual summaries contrasts with current approaches to medical decision support, which mainly rely on visualisation techniques to present data. A recent off-ward evaluation of a prototype system, BT-45, suggested that textual summarisation is at least as effective in supporting decisions as

D. Riaño et al. (Eds.): KR4HC 2009, LNAI 5943, pp. 155–168, 2010.
© Springer-Verlag Berlin Heidelberg 2010

current visualisation techniques [3]. However, the robustness and effectiveness of such systems depend on the extent to which they incorporate a principled approach to temporal representation and uncertainty. To be an effective decision support tool, a summary must permit the reader to reconstruct the temporal sequence of the events that it narrates, and make clear the relations between them. Furthermore, where the precise time at which events occurred is not available, the inference of temporal and/or causal relationships between events can be compromised. The resulting uncertainty should arguably be reflected in the texts produced. Failure to do this can result in erroneous decision-making, whose consequences can be serious given the fault-critical nature of the environment in the NICU.

In this paper, we describe an approach to dealing with temporal uncertainty in the reasoning component of these systems based on Possibility Theory. We also discuss how the outcome of reasoning about temporal relations, and other relations that are contingent upon them, such as causality, can be exploited in an NLG component to communicate uncertainty in the data using expressions such as epistemic modals (e.g. *may* and *must*). The approach is intended to be generalisable to some extent to those kinds of situations in which raw input data needs to be processed prior to carrying out any form of reasoning, and the data itself contains incorrect or uncertain times for the events under consideration. On the other hand, as emphasised above, medical decision support is a particularly important domain in which to deal with these issues, both because of the high density of the data being processed (resulting in increased likelihood of temporal inaccuracy) and the potential consequences of failing to deal with uncertainty.

In the rest of this paper, we first begin by reviewing some related work (Section 2), followed by a motivating example from the NICU domain (Section 3). We then introduce our formalism, with a focus on uncertain relations between intervals (Section 4). Section 5 describes the approach to reasoning with uncertain temporal information. Section 6 discusses how uncertainty in temporal and causal relations can be used to inform the choices made by a text generator in producing a summary, with particular emphasis on the use of modal expressions. Finally, Section 7 reports preliminary results from a web-based experiment testing the association between subjective certainty about a proposition and the participants' choice of linguistic expressions to convey it. We conclude with remarks on future work in Section 8.

2 Related Work

Temporal reasoning is crucial to temporal abstraction [4]. In medicine, it is especially important in making inferences for diagnosis, recommendations based on computerized guidelines, or textual summarisation. Formalisms for temporal reasoning typically rely on the use of temporal constraints [5,6,7,8,9,10] which can either be qualitative (for example, Allen's temporal relations [5]) or numerical [11] (such as a range of temporal distances between two events). Both representations account for different kinds of imprecision: the former is suitable for relations such as *A is after B*; the latter for expressing relations such as *the temporal distance between A and B is between 2 and 4 hours*. With imprecise data and knowledge (which often occurs if the source is a human agent), reasoning leads to uncertainty in temporal relations, which many formalisms represent either through the use of probability distributions [7] or fuzzy sets [6,8].

Regarding probability based reasoning, Ryabov and Trudel [7] have proposed quali-tative probabilistic temporal interval networks in which relations between intervals are labelled with a probability value. While the framework supports reasoning with uncer-tain relations, it presupposes that the probabilities are known. This strong assumption limits the model's applicability in domains where many concepts are to be dealt with (such as the NICU) and precise estimates of probabilities cannot be made due to the absence of large volumes of annotated data. Moreover, probability theory lacks the flexibility to express partial ignorance. Indeed, when the probability of a proposition A is known, the probability of its complement is fully determined ($P(\bar{A}) = 1 - P(A)$). In contrast, non-classical formal theories, such as Possibility Theory or the Dempster-Shafer Theory (DST), deal with partial ignorance by representing uncertainty using two complementary measures (sometimes referred to as upper and lower probabilities).

A recent example of a qualitative temporal reasoning formalism based on fuzzy sets is Badaloni et al.'s IA^{fuz} framework [8], which expresses uncertainty via constraints whose priorities correspond to degrees of plausibility. However, this approach does not address uncertainty derived from data (e.g. inaccurately timestamped events). To deal with vagueness in data, Vila and Godo [12] generalised the classical Temporal Constraint Satisfaction Problem (TCSP) by considering temporal constraints defined by fuzzy sets. This approach has been used in the medical domain for ICU diagnosis [9], and for the processing of clinical texts [10].

More recently, Dubois et al. [6] have proposed a framework for reasoning with a fuzzy version of Allen's temporal relations. This formalism is of linguistic relevance, insofar as there is an intuitive mapping between the representation of graded vague-ness in temporal relations and vague linguistic operators (e.g. *'approximately' equal*), as well as expressions about the certainty that these relations hold (e.g. *A may have happened shortly before B*). Although our focus in this paper is on uncertainty, we are also interested in extending our approach to deal with vagueness (which arises, for ex-ample, in the classification of temporal relations as 'shortly after'). Another attractive property of the formalism is that it derives the uncertainty of temporal relations directly from the intervals over which events occur, by representing these intervals as possibility distributions. This makes the approach amenable to a direct application to raw data. Fur-thermore, Possibility Theory has been shown to better reflect the qualitative nature of human reasoning with uncertainty [13] and is thus better adapted to the communication of uncertainty in natural language for decision making than other formalisms (including DST and other two-level approaches that are exclusively numerical). For these reasons, we build our approach on the basis of this work.

3 A Motivating Example

NICU data is of two kinds: 1) discrete records logged by the medical staff on the NICU database, such as drug administration; 2) physiological data sampled at high frequency from probes measuring heart rate (HR), oxygen saturation (SaO2 or spO2), etc. We as-sume an interval-based representation. The systems under discussion [3] process the input data in four stages: A *data analysis* stage identifies significant trends and pat-terns in the physiological data, as well as data records in the database, mapping them to concepts in a domain-specific ontology. Then, *data interpretation* infers temporal and causal relations between events. Subsequently, the NLG stage selects important events, plans the structure of the summary (*document planning*) and maps the selected events and their relations first to semantic representations (*microplanning*) and finally to syntax (*realisation*). Our focus in this paper is on extensions to data interpretation and microplanning, to deal with uncertainty in temporal relations in reasoning and natural language semantics.

To make the problem concrete, consider the sample of data in Figure 1(a), consisting both of events logged in the database by NICU staff and patterns automatically discovered during signal analysis. A corresponding fragment of a nurse shift summary, written by an experienced neonatal nurse, is shown below.

Example 1. He is currently on nasal CPAP in air, having been extubated today [...] Prior to extubation his SpO2 and HR showed compromise during handling with desaturations and HR decelerations

The example highlights a number of possible sources of uncertainty for a system that (unlike a nurse) is entirely reliant on recorded data. The extubation has not been recorded in the database and must be inferred. There are two relevant facts. First, the baby's having been on SIMV ventilation (a type of ventilation requiring the patient to be intubated, i.e. to have a tube in her throat supplying air) earlier is consistent with her having been intubated. Second, the change to CPAP (a kind of ventilation less severe than SIMV) indicates that the baby has been extubated. The precise time at which the extubation was carried out is, however, uncertain, and there is no specific event corresponding to the placement of the baby on CPAP. The latter too is inferred from the two consecutive ventilator readings, the second of which shows a change in ventilation mode. Thus, the precise time of the ventilator change event itself cannot be determined, though it must have overlapped with the extubation. Finally, the text makes reference to instability in heart rate (HR) and oxygen saturation (SpO2). This is an abstraction that the system needs to perform from the signals. Once again, the interval over which the period of instability holds is fuzzy. Moreover, the reference to 'handling' suggests that the instability occurred during the extubation and prior to its completion.

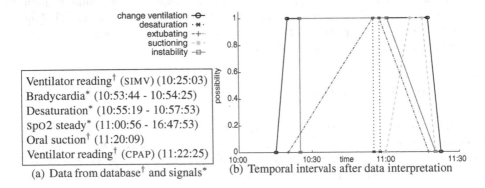

(a) Data from database[†] and signals[*]

(b) Temporal intervals after data interpretation

Fig. 1. Data from database and signal analysis and its temporal representation after interpretation

The outcome of data interpretation is shown in Figure 1(b). The trapezoidal representation of event intervals indicates the inaccuracy in the time at which they started and ended. This affects the certainty with which relationships between them can be inferred, such as the temporal overlap between the ventilator change and the extubation, and the causal relation inferred between the extubation and the instability period, which is supported by the knowledge that handling must have occurred during the extubation, and that it often causes temporary disruptions in a patient's physiological parameters. Though domain knowledge reduces uncertainty, it is not always possible to eliminate it. In such cases, a summary should communicate this uncertainty to the human reader.

4 Representing Temporal Information

The formalism used to represent intervals and uncertain relations between them is built on Possibility Theory as formalised by Dubois *et al.* [6]. In what follows, lowercase italic letters $(a,b,c,...)$ denote dates and normal uppercase (A,B,C,...) inaccurate intervals. Recall that in Possibility Theory, the uncertainty about an inaccurate interval A that holds at date $d \in \mathbf{Z}$ can be evaluated by the dual measures of possibility II and necessity (also called *certainty*) N, as follows:

$$II(A(d)) = h_A(d) \tag{1}$$

$$N(A(d)) = 1 - h_{\bar{A}}(d) \tag{2}$$

Where $h_A \in [0,1]$ is the hold function of the interval A, representing the degree to which A has possibly occurred at date d, and $\bar{A}$ is the complement of A. Additionally, $II(A) = \max h_A$, $II(A(d) \vee B(d)) = \max\{II(A(d)), II(B(d))\}$, $N(A(d) \wedge B(d)) = \min\{N(A(d)), N(B(d))\}$. Moreover, by formula 2 and the following property:

$$II(A \cup \bar{A}) = 1 \tag{3}$$

the necessity of an interval A at date d can be summarised as $A(d)$ *is certain only if no interval contradicting A (i.e., $\bar{A}$) is possible at time d.* If several contradictory intervals are completely possible at the same time (e.g., several mutually exclusive values of a device setting), no certainty exists. In the following, we restrict ourselves to trapezoidal hold functions. Thus, an inaccurate interval is given by the following definition:

Definition 1 (inaccurate interval). *An inaccurate interval A is a 5-tuple* $\langle o, s, e, \alpha, \beta \rangle$, *where* $o \in \mathcal{O}$; $\mathcal{O}$ *is the domain of concepts;* $s, e, \alpha, \beta \in \mathbf{Z}$, $s - \alpha \leq s$, $s \leq e$, *and* $e \leq e + \beta$.

A is seen as a concept with a trapezoidal fuzzy set that describes the period during which A possibly holds. The $[s, e]$ interval is the core of the fuzzy set (i.e. the latest possible start and the earliest possible end) and $[s - \alpha, e + \beta]$ is the support (i.e. the earliest possible start and the latest possible end). In what follows, we reserve the term 'interval' for inaccurate intervals. A trapezoid function leads to $II(A) = 1$, since $\max h_A = 1$. Thus, it is completely possible that A holds, though its start and end time are not known with certainty.

In our application domain, the 'meaning' that an interval conveys can usually be determined by linking each record to a concept in the knowledge representation, which in the present case takes the form of an ontology. One of the uses of associating intervals with concepts is that domain knowledge can then constrain or validate some intervals. A full discussion would take us beyond the scope of this paper, but to summarise, temporal knowledge such as max or min duration or qualitative or quantitative ordering (e.g. A is usually followed by B within a few hours) can be used to remove ambiguity. This kind of expert knowledge is easier to embed into an ontology using fuzzy sets and logic than using probabilistic models. In the following, we distinguish the notion of *event* from the notion of *state*. Typically the former is related to actions or occurrences (e.g. therapy change, bradycardia) over short periods, while the later is related to a conditions which tend to persist over time unless some event perturbs them. Both notions can be represented by an inaccurate interval.

4.1   Temporal Relations between Intervals

Approaches to temporal reasoning [6,7] are usually based on some subset of Allen's 13 relations [5]. Here, we consider only the three disjoint temporal relations *before, intersects, after* (where *intersects* is related to, though different from, Allen's *overlap*) with the aim of dealing with the (un)certainty of temporal relations between intervals (e.g. what is the certainty that A is before B?). Consider the intervals A and B. The necessity that the end of A is before the start of B is given by [6]:

$$N_{es}(A,B) = 1 - \max_{b \leq a \in \mathbf{Z}} \{L(B.s, A.e), \min\{h_A(a), h_B(b)\}\} \tag{4}$$

where $L(x, y) = 1$ if $x \leq y$ and $L(x, y) = 0$ if $x > y$. Thus, the necessity of the end of A occurring before the start of B is the dual of the possibility that the start of B is before the end of A. L is used to constrain $N_{es}(A,B)$ to be 1 when the core of A overlaps the core of B (in which case the possibility that the beginning of B is before the end of A is 1). Similarly, we can define the necessity $N_{ee}(A,B)$ that the end of A is before the end of B, the necessity $N_{ss}(A,B)$ that the start of A is before the start of B, and the necessity $N_{se}(A,B)$ that the start of A is before the end of B. For intervals A and B, we define the three basic relations as follows:

$$N(A \text{ before } B) = N_{es}(A,B) \tag{5}$$

$$N(A \text{ after } B) = N_{es}(B,A) \tag{6}$$

$$N(A \text{ intersects } B) = \min\{N_{se}(B,A), N_{se}(A,B)\} \tag{7}$$

Note that the system described here does not use these relations to maintain a temporal network. Rather, temporal relations are established exclusively on the basis of observation, obviating the need for temporal constraint propagation. This is advantageous, given that the number of intervals can run into thousands, which would compromise the runtime efficiency of such a system were it to attempt to maintain a fully consistent network. Solutions to the latter problem are known to have an exponential worst-case complexity [5]. However, it is precisely because of the reliance on direct observation that errors or uncertainties in the input data need to be rectified at the reasoning stage.

5 Temporal Abstraction and Interpretation

Interpretation is done in two stages: 1) application of *a priori* domain knowledge to independent intervals to alter their fuzzy sets to reduce ambiguity; and 2) application of temporal reasoning to abstract and interpret states and events. Our knowledge base consists of a large ontology developed within the BabyTalk project [3] (which contains more than 900 concepts) and rules acquired from interviews with experts.

For the first phase, when the intervals are first read in from the data source, knowledge can be deployed to directly constrain uncertainty about their temporal information. For example, the ventilator mode readings (i.e., SIMV and CPAP) in Figure 1 are not recorded with accurate timestamps (they are logged on an hourly basis). Due to property (3), the possibility of each of the ventilation mode values should be defined between these two readings. Theoretically, it could take any value in its domain. Since any value for the ventilator reading is possible at any time in principle, no certainty of any kind can be derived from these data alone, without reasoning based on the application of reasonable *a priori* constraints. These constraints consist in:

1. assuming persistence for states, unless there is evidence that a state has ceased to hold;
2. assuming that any state A is expanded by *delay*, to take into account a minimal delay between the human observation and the human recording on the computer;
3. assuming that any event A has actually happened before the transaction date (that is, the date at which the record for A is entered).

These simple rules are crucial for disambiguation. The first constraint enables the aggregation of intervals representing states with the same properties. The second one accounts for inertia in the recording of data (i.e. a value is only recorded if it has held true for a certain delay period around the transaction date). The third one is known to be typically true based on consultation with domain experts. Apart from these rules, knowledge encoded in the ontology (such as max and min duration) and in the expert rules, permits the modification of the fuzzy set of the interval. For the example in Figure 1(b), the baby is known to be intubated; however, the CPAP reading contradicts this, since the domain knowledge specifies that this kind of ventilation support *requires* the baby to not be intubated. Thus, the system infers that the ventilation mode has been changed over the period C and that an extubation event E has possibly existed between the two readings. The exact location of this extubation is still vague but again the knowledge base informs us that extubation can cause perturbation on the physiological signals (due, among other things, to handling). Thus, the maximal possibility for this extubation occurs during the period of the desaturation D, which is intersected by E. This reasoning explains the shape of E in Figure 1(b), which is completely possible during D, less possible during the change of ventilation period, and impossible otherwise. These outcomes are clearly strongly dependent on domain assumptions, but this cannot be neglected, as basic knowledge of the domain is often an easy and reliable way to reduce complexity in reasoning.

After the first stage, the following values can be computed: N(C after E) $= 0$, and N(C intersects E) $= 1$. For E and the oral suction O, N(E before O) $= 0.42$, N(E after O) $= 0$, N(E intersects O) $= 0.58$. In addition, the following relationships are computed between the oral suction event O and the period of instability U in HR and spo2: N(O intersects U) $= 0.68$, N(O after U) $= 0.32$.

Thus, although the temporal order of oral suction and extubation could not be established, it is more certain that suction has been performed right after or during extubation than before. This also explains part of the motivation behind the suction event.

Finally, inference rules are applied for abstraction and interpretation. In this framework, the validity of an inference chain is measured by its weakest links, so that the weight of a conclusion should be the weakest among the weights of its premises. As noted earlier, our reasoning is data-driven in the sense that the temporal relations considered are all and only those derived from data. As an example, the following rule is fired when an intervention (such as an extubation E or an oral suction O) intersects with an instability period (U), which represents the degree of variation of the physiological parameters related to respiration over periods of time. These periods are delimited by the main respiratory interventions.

Example 2. E is-a RESPIRATORY INTERVENTION $\wedge$ U is-a INSTABILITY $\wedge N$(E intersects U) $\geq \psi \Rightarrow N$(E causes U) $= N$(E intersects U)

This rule matches the extubation and the oral suction in the example and infers that *O causes U* with necessity 0.68, while *E causes U* with necessity 1.

6   From Events to Text

In this section, we shall be concerned with the implications of the foregoing discussion for communicating uncertainty in the microplanning component. This takes as input a document plan — a labelled graph whose nodes are intervals or sequences of events, and whose edges are relations between intervals — and produces a semantic representation for the intervals in the document plan, which is then mapped to a syntactic representation. A partial document plan for our example is displayed in Figure 2, where the edge labels indicate the necessity with which a relation holds.

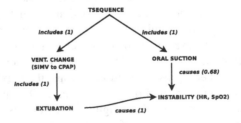

Fig. 2. Document plan fragment. Numbers in parentheses are necessity values.

One of the mechanisms that natural language provides for the expression of different degrees of (un)certainty is modality. Classical treatments of the semantics of modal expressions such as *can*, *must* and *may*, rely on their *modal force* (the degree of necessity or possibility expressed by the modal expression) and the *contextual background* against which they are interpreted. In the present case, our focus shall be on epistemic modality, where the relevant background is the speaker's knowledge. A proposition such as *x must/may have occurred* is roughly paraphrasable as *x must/may have occurred* in view of what is known [14]. Assuming, following Grice [15], that a speaker will not impart information beyond what is required unless it is relevant, qualifying an assertion in this way (e.g. *The extubation may have caused instability*) signals to the hearer/reader that the degree of a speaker's certainty is relevant to how the truth of the proposition should be evaluated [16]. This is particularly relevant for the present domain, where expressing uncertainty explicitly may alter the course of decision-making by a reader. It seems likely that this is also true of *must*. Although this has traditionally been taken as expressing logical necessity [14], the use of *must* suggests that the relativisation to the speaker's knowledge is important; this seems to be part of the pragmatic import of the use of the modal, and is compatible with the Gricean argument outlined above.

Typically, formal semantic treatments of modals are couched in a possible worlds framework [14,16]; this has also been adopted in NLG by Klabunde [17] to deal with (deontic) modality in a system that generates recommendations for course choices to students. In contrast to this work, the present approach proposes to view epistemic modal expressions as involving a direct mapping from different degrees of necessity or certainty (which reflects the epistemic 'modal force' of the proposition to be communicated) to linguistic expressions. One potential advantage of this approach is that, just as the necessity and possibility computations discussed in Section 4 are derived directly from data, so the use of epistemic modals is grounded in the data that constitutes the speaker's (system's) knowledge state.

To deal with modality in this way, we make the following assumptions about the lexical resources available to the microplanner. First, every relation between intervals in a document plan maps to a linguistic expression in the lexicon. For example, a *cause* relation maps to the verb *cause*. Modal auxiliaries are represented in the lexicon via a function $\mu : N(R) \rightarrow$ AUX, which maps the necessity value of a relation to an epistemic modal auxiliary verb. A possible implementation of this function is sketched out below:

$$\mu(N(R)) = \begin{cases} may \text{ if } l < N(R) \leq 0.6 \\ must \text{ if } 0.6 < N(R) < 1 \\ \bot \text{ otherwise.} \end{cases} \tag{8}$$

where l is a lower bound on the certainty below which the relation is not expressed at all because it is too uncertain, and $\bot$ is `null`. By this formulation, a relation such as *cause* will be expressed with no qualification ($\bot$) if the certainty is 1, but may be qualified using *may* (which carries weak epistemic modal force) or *must* otherwise. Another possibility for expressing high degrees of certainty is *should*. However, a sentence such as *X should have caused Y* may be interpreted as implying violated expectation (i.e., *X was expected to have caused Y but didn't*). If this is the case, then using *should* would take the system's generated text beyond description and into something akin to recommendation, since pointing out violated expectation may lead to an increased focus on the reader's part to check whether something went wrong.

The document plan in Figure 2 contains an additional complication: there are two events possibly contributing to the instability event, with different degrees of certainty. Here, there are two possibilities. If the extubation and suction events are aggregated, to form a single clause as in Example 3, then the certainty of their joint causal role in producing the instability is once again the weakest link in the causal chain (the minimum certainty value), following the reasoning adopted in Section 5.

Example 3. The baby was moved from SIMV to CPAP. He was extubated and underwent oral suction. This must have caused the instability in HR and SpO2.

An alternative strategy is to realise each clause separately, making the causal link explicit in each case. In the case of extubation, where the certainty is 1, no modal is used. In the second case, the assertion of causality is qualified via *must*.

Example 4. The baby was moved from SIMV to CPAP. He was extubated, causing the instability in HR and SpO2. He underwent suction. This too must have caused instability.

The best choice between these two alternatives is an open empirical question.

7 Empirically Grounding the Linguistic Model

The above illustration of how necessity values can inform lexical choice in microplanning is couched in largely intuitive terms. However, it throws up a number of questions which we are currently investigating. Among the relevant issues is the degree of certainty with which people interpret different modal expressions in epistemic contexts, as well as the other inferences that they generate. An answer to this question would serve as the basis for empirically grounding the lexical resources used by the system, as well as testing our intuitions regarding the different modal force of different expressions.

In order to answer these questions, we ran an experiment aimed at investigating the degree of certainty with which propositions describing simple events are interpreted by human speakers depending on the degree of temporal uncertainty associated with the events. Another aim of the experiment was to investigate speakers' choice of linguistic expressions to express uncertainty, with a view to incorporating this into our model of lexical choice. The linguistic expressions considered fall into three classes: 1) Epistemic modals (*must* and *may*), which are the focus of the previous section; 2) Adverbs of possibility (*possibly* and *perhaps*), which offer an alternative way of expressing uncertainty and were included for comparison; and 3) Negation (that is, sentences of the form *it is not the case that E occurred at t*).

7.1 Materials, Design and Procedure

The experiment was conducted over the web. Participants were shown a series of scenarios, each of which consisted of a background text and two temporally grounded propositions (**S1** and **S2**) describing two events (**E1** and **E2**), whose timing could be precisely or inaccurately known (see the top of Figure 3). The scenarios were designed

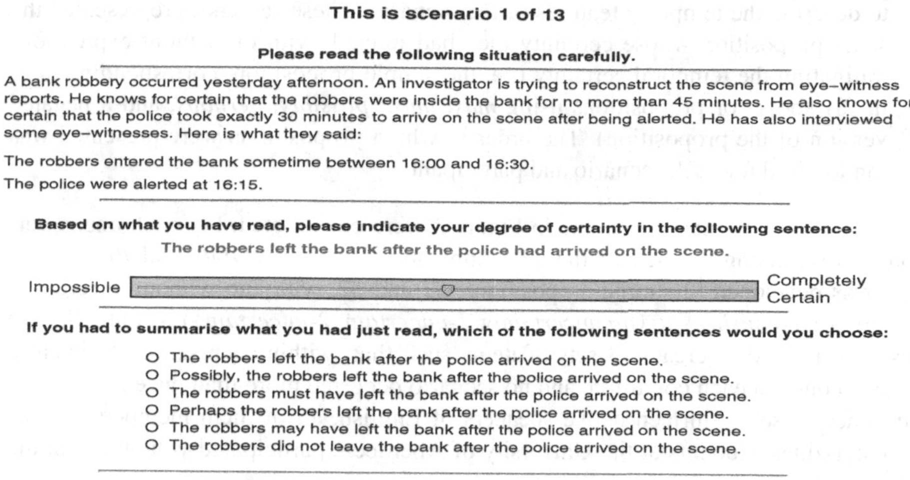

Please read the following situation carefully.

A bank robbery occurred yesterday afternoon. An investigator is trying to reconstruct the scene from eye–witness reports. He knows for certain that the robbers were inside the bank for no more than 45 minutes. He also knows for certain that the police took exactly 30 minutes to arrive on the scene after being alerted. He has also interviewed some eye–witnesses. Here is what they said:

The robbers entered the bank sometime between 16:00 and 16:30.
The police were alerted at 16:15.

Based on what you have read, please indicate your degree of certainty in the following sentence:

The robbers left the bank after the police had arrived on the scene.

Impossible |═══════════════════════════════▽═══════════════════════| Completely Certain

If you had to summarise what you had just read, which of the following sentences would you choose:

- ○ The robbers left the bank after the police arrived on the scene.
- ○ Possibly, the robbers left the bank after the police arrived on the scene.
- ○ The robbers must have left the bank after the police arrived on the scene.
- ○ Perhaps the robbers left the bank after the police arrived on the scene.
- ○ The robbers may have left the bank after the police arrived on the scene.
- ○ The robbers did not leave the bank after the police arrived on the scene.

Fig. 3. Screenshot of one of the thirteen scenarios shown to the participants sequentially

to make it explicit that the events themselves actually happened for certain and that uncertainty was only related to their timing.

The experiment manipulated two factors. *Uncertainty* (3 levels) manipulated the extent to which the two events were precisely located in time. In the *no uncertainty* case, event times were expressed with a crisp value (e.g., *The robbers entered the bank* at 16:00.); in the *1-uncertainty* case, **E1** was expressed with a fuzzy temporal interval (e.g., *The robbers entered the bank* sometime between 16:00 and 16:30.); in the *2-uncertainty* case, both events had fuzzy temporal intervals. This factor enabled us to control the degree of certainty of temporal relations between events. The second factor, *Proposition Type* (4 levels), manipulated the type of proposition whose subjective certainty participants were asked to judge, namely: a simple proposition describing either **E1** or **E2**; or a compound proposition describing the temporal relation between the two events using one of the temporal connectives *before, after,* or *during*. Once participants had read a background text, they were asked to perform two tasks:

- **Judgement:** Given a certain scenario involving two events, participants were asked to judge their certainty that an event happened at a certain time or in a certain temporal order in relation to another event. Certainty was judged using a slider (see Figure 3 middle) representing the 'Ψ-scale' [13], which combines both possibility and necessity measures and ranges from 'impossible' to 'completely certain'. From the Ψ measure, the corresponding possibility II and necessity N can easily be reconstructed using (9).

$$II(P) = \begin{cases} 2 * \Psi \text{ if } \Psi \leq 0.5 \\ 1 \quad\quad \text{if } \Psi > 0.5 \end{cases}, \ N(P) = \begin{cases} 0 \quad\quad\quad \text{if } \Psi \leq 0.5 \\ 2 * \Psi - 1 \text{ if } \Psi > 0.5 \end{cases} \quad (9)$$

- **Forced choice:** After judging the certainty of the proposition, they were asked to select, from among a set of sentences, the one they thought was most appropriate to describe the temporal features of the scenario. These sentences represented the same proposition whose certainty they had judged, with or without expressions mitigating the temporal certainty (i.e. the simple proposition, corresponding to $\perp$ in (8) above, or propositions using *may*, *must*, *possibly*, *perhaps*, and a negated version of the proposition). The order in which propositions were presented was randomised for each scenario and participant.

Thirteen scenarios such as the one in Figure 3 were constructed. For each one, a version corresponding to each of the 13 combinations of *Uncertainty* and *Proposition Type* was developed (2 simple propositions, *S1* and *S2*, with and without uncertainty + {*before,after,during*} × {*no uncertainty, 1-uncertain, 2-uncertain*}). A Latin Square Design was used to create 13 sets of items such that, within each set, each scenario occurred once in each condition, and no scenario occurred more than once in that condition across sets. Thirteen native speakers of English, all of them members of staff or postgraduate students at the University of Aberdeen, participated voluntarily in the experiment.

7.2 Results

Separate univariate ANOVAs were conducted to test the effect of *Uncertainty* and *Proposition Type* on the Ψ score, as well as on the possibility and necessity values. There was a significant main effect of *Uncertainty* on both Ψ ($F(2, 158) = 8.657; p < .001$) and on the derived necessity values ($F(2, 158) = 20.145; p < .001$), but not on possibility ($F(2, 158) = 0.003; ns$). Proposition Type exerted no main effect and there was no interaction between the two factors. This suggests that judgements of subjective certainty are strongly influenced by the manipulation of temporal uncertainty in the propositions being judged; however, it is *necessity* rather than possibility, which is the key correlate of these judgements. Thus, our focus on necessity values in the linguistic model sketched out in Section 6 has some prima facie justification.

Table 1 displays the mean Ψ, possibility and necessity values associated with the different types of propositions, based on participants' choices. In the default ($\perp$) case, both necessity and possibility are high. This is expected, given that, in natural language, unqualified assertions tend to be made in case subjective certainty is high. For *may* and *must*, the associated possibility values are high in both cases. On the other hand, there is a clear difference between sentences qualified with such modals and the default case: the latter is associated with a higher necessity value. Moreover, *may* involves lower necessity than *must*, as expected. As for adverbials and negation, necessity tends to be much lower, but the differences in possibility are clearer than in the case of modals.

A stepwise multinomial regression analysis to test the significance of II and N in determining the category of phrase selected by participants revealed that both II and N play a significant role in determining phrase choice (II: $\chi^2 = 61.8; p < .001$; N: $\chi^2 = 53.89; p < .001$) . Given the differences in the mean II and N values for modals on the one hand, and adverbials and negation on the other, separate regression analyses were conducted to identify the role of the two measures on the choice within either class. Interestingly, the models suggest a dissociation between necessity and possibility. Within the class of the two modals (together with the default $\perp$ case), regression showed

Table 1. Mean and standard deviations for subjective certainty (Ψ) and corresponding necessity and possibility values, as a function of phrase choice

	$\perp$	must	may	perhaps	possibly	negation
Ψ	.94 (.1)	.86 (.17)	.49 (.15)	.59 (.13)	.48 (.23)	.13 (.25)
possibility (Π)	1 (0)	1 (.01)	.89 (.19)	.97 (.06)	.83 (.33)	.21 (.35)
necessity (N)	.88 (.2)	.71 (.34)	.09 (.19)	.21 (.23)	.13 (.24)	.05 (.20)

a significant effect of necessity ($\chi^2 = 36.07; p < .001$), but not possibility ($\chi^2 = 4.3; p > .1$). In contrast, the choice between the two adverbials and negation showed a significant role of Π ($\chi^2 = 28.31; p < .001$) but not N ($\chi^2 = 1.12; p > .5$).

To summarise, the judgement results indicate that necessity is the primary correlate of people's subjective certainty judgements, but the phrase choice data suggests a dual role for possibility and necessity values. Specifically, there seems to be a dissociation between epistemic modals on the one hand, and adverbials of possibility (and negation) on the other. One possibility is that these linguistic expressions express different factors contributing to overall subjective certainty.

Although the results suggest that the intuitions underlying the linguistic model presented in Section 6 are on the right track, the mapping from necessity values to modals can be fine-tuned on the basis of this data. Subjects seemed to be tolerant of *some* degree of subjective uncertainty in opting for a non-qualified utterance ($\perp$), whereas our original proposal was to use $\perp$ only when $N = 1$. On the other hand, *must* is selected in cases where subjective certainty is quite high, and the gap between *must* and *may* is substantial. This still leaves open the question of the role of possibility, particularly of its apparent predictive power for the choice of adverbials and/or negation.

Though they are encouraging, the above results are preliminary, both in the sense that they are based on a relatively small pool of 13 subjects, and because they afford more sophisticated analysis. In addition to gathering more data, our ongoing work is addressing the question whether the subjective (un)certainty of two propositions linked by a temporal relation can be predicted from that of the simple propositions.

8   Conclusions and Future Work

This paper has proposed a possibility theoretic approach to the representation of inaccurate intervals and the knowledge-based discovery of temporal relations. The expression of these in text generation uses the uncertainty measure for qualifying relations via epistemic modal expressions. Our formalism is attractive in that it can be used to combine information from different sources (e.g., pattern recognition outputs, database entries, information extracted from free-text) while linguistic expressions are also directly grounded in measures of certainty based on the available information. Preliminary experimental work suggests that the model is on the right track, though the range of expression types could be expanded beyond modals, by taking into account a possible difference between the linguistic expression of necessity and possibility. Further analysis of our experimental data should shed further light on the differences between linguistic expressions of (un)certainty and their interpretation, as well as the relationship between adverbials and epistemic modals. In addition, subjective certainty judgements can be used to empirically validate our temporal reasoning model, by comparing the model prediction of certainty of a temporal relation from the certainty of its component events, to the actual values rated by subjects. Finally, our experiment also needs to be extended to more specialised (and more fault-critical) cases, such as that of the NICU.

Acknowledgements

Thanks to Jim Hunter, Ehud Reiter, Kees van Deemter and Neil McIntosh for helpful comments on this work. We also thank the referees for their comments. Part of this work was supported by UK EPSRC grants EP/D049520/1 and EP/D05057X/1.

References

1. Terenziani, P., Snodgrass, R.T., Bottrighi, A., Torchio, M., Molino, G.: Extending temporal databases to deal with telic/atelic medical data. In: Miksch, S., Hunter, J., Keravnou, E.T. (eds.) AIME 2005. LNCS (LNAI), vol. 3581, pp. 58–66. Springer, Heidelberg (2005)
2. Barnett, G., Famiglietti, K., Kim, R., Hoffer, E., Feldman, M.: DXplain on the internet. In: Proceedings of AMIA1998, pp. 607–611 (1998)
3. Gatt, A., Portet, F., Reiter, E., Hunter, J., Mahamood, S., Moncur, W., Sripada, S.: From data to text in the neonatal intensive care unit: Using NLG technology for decision support and information management. AI Communications 22, 153–186 (2009)
4. Stacey, M., McGregor, C.: Temporal abstraction in intelligent clinical data analysis: A survey. Artificial Intelligence in Medicine 39(1), 1–24 (2007)
5. Allen, J.: Maintaining knowledge about temporal intervals. Communications of the ACM 26(11), 832–843 (1983)
6. Dubois, D., Allel, H., Prade, H.: Fuzziness and uncertainty in temporal reasoning. Journal of Universal Computer Science 9(9), 1168–1194 (2003)
7. Ryabov, V., Trudel, A.: Probabilistic temporal interval networks. In: Proceedings of TIME 2004, pp. 64–67 (2004)
8. Badaloni, S., Giacomin, M.: The algebra IA^{fuz}: a framework for qualitative fuzzy temporal reasoning. Artificial Intelligence 170(10), 872–908 (2006)
9. Palma, J., Juareza, J.M., Camposa, M., Marina, R.: Fuzzy theory approach for temporal model-based diagnosis: An application to medical domains. Artificial Intelligence in Medicine 38(2), 197–218 (2006)
10. Lai, A.M., Parsons, S., Hripcsak, G.: Fuzzy temporal constraint networks for clinical information. In: Proceedings of AMIA 2008, pp. 374–378 (2008)
11. Dechter, R., Meiri, I., Pearl, J.: Temporal constraint networks. Artificial Intelligence 49(1-3), 61–95 (1991)
12. Vila, L., Godo, L.: On fuzzy temporal constraint networks. Mathware & soft computing 1(3), 315–334 (1994)
13. Raufaste, E., da Silva Neves, R., Mariné, E.: Testing the descriptive validity of possibility theory in human judgements of uncertainty. Artificial Intelligence 148, 197–218 (2003)
14. Kratzer, A.: What *must* and *can* must and can mean. Linguistics and Philosophy 1, 337–355 (1977)
15. Grice, H.: Logic and conversation. In: Cole, P., Morgan, J. (eds.) Syntax and Semantics: Speech Acts. Academic Press, London (1975)
16. Papafragou, A.: Epistemic modality and truth conditions. Lingua 116, 1688–1702 (2006)
17. Klabunde, R.: Lexical choice for modal expressions. In: Proceedings of ENLG 2007 (2007)

Argumentation about Treatment Efficacy

Nikos Gorogiannis[1], Anthony Hunter[1], Vivek Patkar[2], and Matthew Williams[3]

[1] Department of Computer Science, University College London,
Gower Street, WC1E 6BT, London, UK
{n.gkorogiannis,a.hunter}@cs.ucl.ac.uk
[2] University College London Cancer Institute,
72 Huntley Street, London WC1E 6BT, UK
v.patkar@medsch.ucl.ac.uk
[3] Mount Vernon Hospital, Northwood, Middlesex, HA6 2RN, UK
matt.williams@nhs.net

Abstract. The volume and complexity of knowledge produced by medical research calls for the development of technology for automated management and analysis of such knowledge. In this paper, we identify scenarios where a researcher or a clinician may wish to use automated systems for analysing knowledge from clinical trials. For this, we propose a language for encoding, capturing and synthesising knowledge from clinical trials and a framework that allows the construction of arguments from such knowledge. We develop this framework and demonstrate its use on a case study regarding chemotherapy regimens for ovarian cancer.

1 Introduction

During the last few decades, medical research has grown rapidly, producing an enormous amount of results. This growth necessitates a change in how this knowledge is managed, searched and analysed in order to achieve the full potential offered. Information retrieval techniques are already indispensable for users. However, they do not address all the needs, as they do not enable one to *reason* with the available knowledge. This is where knowledge representation techniques can make a contribution. In this paper, we propose a framework for expressing and analysing this knowledge in an effective manner.

One important kind of such knowledge is associated with superiority-testing clinical trials, a type of clinical study which involves comparing the efficacy of two or more treatments when given to a particular class of patients. Each such trial reports the results of comparing two or more treatments. In order to have a global view of the relative comparisons between treatments for a particular condition, a potentially large number of publications needs to be reviewed and this is why syntheses of this kind of knowledge are undertaken. Normally, a group of scientists would search the literature and perform a study employing background medical knowledge and statistical techniques for aggregating clinical trial results (e.g., a systematic review or meta-analysis), requiring significant time and effort. In addition, such syntheses of the literature can quickly become out of date if new results are published in the interim.

D. Riaño et al. (Eds.): KR4HC 2009, LNAI 5943, pp. 169–179, 2010.

Therefore, getting a quick, up-to-date review of the state of the art on treatment efficacy for a particular condition is not always feasible. Thus, it would be helpful to have a method for automatically analysing and presenting the clinical trial results and the possible ways to aggregate those in an intuitive form, highlighting agreement and conflict present within the literature. Our proposal aims to suggest such a method. The first part of the proposal is a language that can be used to encode the published results in a semantically appropriate way, and methods for constructing a knowledge base from the encoded results. Using such a knowledge base clinical scientists can access easily the desired set of evidence. The second part in our framework allows the construction of arguments on the basis of evidence as well as their syntheses, published or generated on-the-fly. These arguments encode judgments about the comparison of efficacy of two treatments, thus representing interpretations of the available evidence. These are then presented and organised according to the agreement and conflict inherent. In addition, users can encode preferences for automatically ruling in favour of the preferred arguments in a conflict.

To demonstrate our proposal, we present worked examples from a case study on chemotherapy treatments for ovarian cancer. We use an existing paper that reports the results of synthesising the available evidence in the area [1], and show how the user would navigate the evidence in the literature, create his/her own syntheses and produce arguments and counter-arguments for comparisons of treatment efficacy.

2　A Language for Representing Clinical Trial Results

Our focus will be on 2-arm superiority trials, i.e., clinical trials whose purpose is to determine whether, given two treatments, one is superior to the other (strictly speaking, such a trial tries to disprove the hypothesis that the two treatments are identical). This is an extremely common trial design; other trial design types do exist, but we will ignore them here for simplicity.

We give below the details of two results used in [1] (which we will call the Ovarian Cancer Study from now on), encoded in the proposed language, which will be explained in this section.

Result	tr_1	tr_2	ind	hr	sig
γ_{Markman}	$PTC\bar{I}$	$PTCI$	S	1.23	$\bot$
$\gamma_{\text{Armstrong}}$	$PTC\bar{I}$	$PTCI$	S	1.33	$\top$

The first row corresponds to a superiority trial performed by Markman et al in 2001, which compared two chemotherapy regimens for ovarian cancer patients: the combination of agents paclitaxel and cisplatin, administered intravenously (IV) with the combination of carboplatin (IV), paclitaxel (IV) and carboplatin, administered intraperitoneally (IP). The study produced a hazard ratio of 1.23 for survival and did not find a statistically significant difference between the two regimens. We now explain what these data mean and how we encode them.

To represent a trial result we will use γ, possibly with subscripts, e.g., γ_{Markman} for the trial result by Markman et al. We also will use Γ to denote the set of results under consideration. The characteristics of a trial result will be called **attributes**. We use a set *Attr* of partial functions from trial results to values, to represent the set of available characteristics.

The first attribute of interest of a trial is the patient class involved. *Medical ontologies* provide languages for capturing patient characteristics (among other things) as well as logical machinery for answering queries about them. Description logics are usually employed to provide the necessary inference tools (see [2]). Ontologies, however, fall outside the scope of this paper; for simplicity we assume that the set of results Γ concerns a particular, sensible patient class.

The next component of our language concerns treatments. Again, medical ontologies cater for this task by providing categories and relationships on treatments, substances used, and other characteristics. For simplicity, we will elide such complexity and assume that there is a set algebra of treatments $\mathcal{T}$ whose members, denoted by τ possibly with subscripts, represent specific treatments. In the example, the phrases "paclitaxel and cisplatin, administered intravenously (IV)" and "carboplatin (IV), paclitaxel (IV) and carboplatin, administered intraperitoneally (IP)" would be denoted by, e.g, τ_1 and τ_2. We will use $\text{tr}_1(\gamma)$ and $\text{tr}_2(\gamma)$ to signify the treatments compared in the trial result γ.

Clinicians and statisticians, however, may want to group treatments together in order to allow for more inferential power. In such cases, it is customary to conflate the details of the treatments down to the level of abstraction required. For example, the Ovarian Cancer Study is only concerned with four characteristics of the treatment: whether or not it employs platinum-based agents (P), taxanes (T), whether it combines several agents (C), and whether it employs at least one agent with intraperitoneal administration (I). We will use four symbols to denote each class of treatments, with a bar over a symbol to denote its negation. Therefore, τ_1 is a member of the treatment group $PTC\bar{I}$ and τ_2 is a member of $PTCI$. From now on we use this treatment clustering scheme for defining $\mathcal{T}$ and assume that tr_1, tr_2 will range over the treatment groups allowed. We also use the term *treatment* to refer to any treatment group.

A trial comparing two treatments will do so with respect to particular outcomes, e.g., in the case of the Ovarian Cancer Study, survival. Moreover, a specific statistical measure will be used to evaluate outcomes. In our example, the hazard ratio for survival is used, which we shall explain below. We call the particular way an outcome is represented in a trial result, the **outcome indicator** and use the attribute ind to retrieve it, given a particular trial. Hence, in our example, $\text{ind}(\gamma_{\text{Markman}}) = S$, denoting hazard ratios for survival. Here, for simplicity, we only consider sets of results that involve the same outcome indicator.

Finally, a trial would report the results of the statistical method comparing the two treatments. There are many different measures, each appropriate to specific trial designs and outcomes. The trials examined in the Ovarian Cancer Study use *hazard ratios*. For a given trial it is possible, using statistical methods which are outside the scope of this paper, to estimate the number of expected events (death in the case of survival) in each group on the basis of the assumption that the two treatments have identical effects (null hypothesis). Then, the *hazard rate* for each group can be computed, which is the ratio of the number of observed events to the number of expected events. The hazard ratio is the ratio of the hazard rates for the two groups. In the case of survival, a hazard ratio over 1 indicates that the group treated with τ_1 fared worse than the group treated with τ_2. We will use the function $hr \in Attr$ applied to the trial result $\gamma \in \Gamma$ to retrieve the value of the hazard ratio reported, $hr(\gamma)$. In addition, a trial will report an indicator of whether this comparison is statistically significant. This can be done in various ways, e.g., whether the p-value is greater or lower than the significance level (commonly 5%), or whether the confidence interval includes the hazard ratio of 1 (in which case the possibility that the two treatments have exactly the same effect cannot be ruled out). However this is reported, the aim is to ascertain whether the result is statistically significant, so we will use a boolean attribute $sig(\gamma)$ to denote whether a result γ is statistically significant. Therefore, for our example we would have $hr(\gamma_{\mathrm{Markman}}) = 1.23$, and $sig(\gamma_{\mathrm{Markman}}) = \bot$.

3   Arguments Drawn from Clinical Trial Results

For a superiority clinical trial comparing treatments τ_1 and τ_2 with respect to the outcome indicator I, there are four possible interpretations of its results:

1. $\tau_1 >_I \tau_2$, meaning that we believe that the result supports the inference that treatment τ_1 is superior to τ_2 with respect to I.
2. $\tau_1 <_I \tau_2$, as above.
3. $\tau_1 \sim_I \tau_2$, meaning that we interpret the result as supporting the inference that neither τ_1 nor τ_2 is superior to each other with respect to I.
4. Finally, it may be the case that we believe that the result does not support any of the above possibilities.

We will call the formulae employed in statements 1–3 **claims**. Formally, any formula of the form $\tau_1 >_I \tau_2$, $\tau_1 \sim_I \tau_2$ and $\tau_1 <_I \tau_2$ will be called a claim, denoted by ϵ, possibly subscripted. We will sometimes use $\circ, \diamond$ as meta-variables for the symbols $>$, $<$ and $\sim$. So, for example, $\tau_1 \circ_I \tau_2$ will stand for either of: $\tau_1 >_I \tau_2$, $\tau_1 <_I \tau_2$ or $\tau_1 \sim_I \tau_2$.

Note that the interpretation of the results of a trial is a very complex and, in some cases, open question and that conflicting answers exist. For example, a strict statistical interpretation would be that a superiority trial can never provide evidence supporting the equivalence of two treatments, *by design*. However, some clinicians would argue that a sufficiently large superiority trial that fails to show superiority is, in fact, in itself evidence of (rough) equivalence. We will refrain from debating these issues, as they are well outside the scope of this work. However, we will allow for different users to express their own semantics.

Given a set of results Γ one can informally think of an argument comprising of a set of evidence, an inferential rule and a conclusion or claim. For example, a plausible interpretation of $\gamma_{\mathrm{Armstrong}}$ is that since $hr(\gamma_{\mathrm{Armstrong}}) > 1$, it indicates that the first treatment is worse than the second with respect to S, i.e., that $PTC\bar{I} <_S PTCI$. We define this process of inference as an inference rule.

Definition 1. *An **inference rule** is a set of conditions (employing set-theoretic expressions and equations utilising attributes over the reals) on a set of results $X \subseteq \Gamma$ and a claim ϵ.*

From now on we will use λ, possibly subscripted, to denote an inference rule, and Λ for the set of all inference rules chosen by the user.

Example 1. One of the simplest inference rules interprets a statistically significant trial result as evidence of actual treatment superiority.

λ_s: Let $X = \{\gamma\}$. If $\mathsf{sig}(\gamma) = \top$ then:
- if $\mathsf{hr}(\gamma) < 1$ then $\mathsf{tr}_1(\gamma) >_{\mathsf{ind}(\gamma)} \mathsf{tr}_2(\gamma)$,
- if $\mathsf{hr}(\gamma) > 1$ then $\mathsf{tr}_1(\gamma) <_{\mathsf{ind}(\gamma)} \mathsf{tr}_2(\gamma)$.

Example 2. Conversely, a non-significant result can be viewed as saying that neither treatment is better than the other one in terms of the outcome indicator considered. This can be captured by the following inference rule:

λ_n: Let $X = \{\gamma\}$. If $\mathsf{sig}(\gamma) = \bot$ then $\mathsf{tr}_1(\gamma) \sim_{\mathsf{ind}(\gamma)} \mathsf{tr}_2(\gamma)$.

An inference rule can be thought of as an *argument generator* in that, on the basis of a set of results that stands as evidence, it supports a certain claim. Therefore, we will define an argument as an application of an inference rule.

Definition 2. *An **argument** is a triple $\langle X, \lambda, \epsilon \rangle$ where $X \subseteq \Gamma$ is a set of results, λ is an inference rule, X satisfies the conditions of λ and ϵ is the claim of λ applied to X.*

Example 3. Given $\gamma_{\mathrm{Armstrong}}$, the following tuple is an argument:

$$A_1 = \langle \{\gamma_{\mathrm{Armstrong}}\}, \lambda_s, PTC\bar{I} <_S PTCI \rangle.$$

By using λ_n on $\gamma_{\mathrm{Markman}}$ we obtain the following argument.

$$A_2 = \langle \{\gamma_{\mathrm{Markman}}\}, \lambda_n, PTC\bar{I} \sim_S PTCI \rangle$$

Clearly, A_1 and A_2 are conflicting. We focus on conflict in the next section.

Definition 3. *Given a claim ϵ, we define the **set of arguments relevant to** ϵ as*

$$\mathsf{args}(\epsilon) = \{\, A \mid A = \langle X, \lambda, \epsilon \rangle \text{ is an argument with } X \subseteq \Gamma, \lambda \in \Lambda \,\}.$$

Given a pair of treatments τ_1, τ_2 and an outcome indicator I we define the set of arguments relevant to τ_1, τ_2 with respect to I as

$$\mathsf{args}(\tau_1, \tau_2, I) = \mathsf{args}(\tau_1 >_I \tau_2) \cup \mathsf{args}(\tau_1 \sim_I \tau_2) \cup \mathsf{args}(\tau_1 <_I \tau_2)$$

Example 4. Let $\Gamma = \{\gamma_{\text{Markman}}, \gamma_{\text{Armstrong}}\}$. Then,

$$\text{args}(PTC\bar{I} <_S PTCI) = \{\langle \{\gamma_{\text{Armstrong}}\}, \lambda_s, PTC\bar{I} <_S PTCI\rangle\}$$
$$\text{args}(PTC\bar{I} \sim_S PTCI) = \{\langle \{\gamma_{\text{Markman}}\}, \lambda_n, PTC\bar{I} \sim_S PTCI\rangle\}$$
$$\text{args}(PTC\bar{I} >_S PTCI) = \emptyset$$
$$\text{args}(PTC\bar{I}, PTCI, S) = \left\{ \begin{array}{l} \langle \{\gamma_{\text{Armstrong}}\}, \lambda_s, PTC\bar{I} <_S PTCI\rangle, \\ \langle \{\gamma_{\text{Markman}}\}, \lambda_n, PTC\bar{I} \sim_S PTCI\rangle \end{array} \right\}$$

The intention behind these definitions is that users should be able to define or select their own inference rules for argument construction.

4   Preferences over Arguments

In Example 4 we saw two arguments that were clearly in conflict:

$$\langle \{\gamma_{\text{Armstrong}}\}, \lambda_s, PTC\bar{I} <_S PTCI\rangle, \langle \{\gamma_{\text{Markman}}\}, \lambda_n, PTC\bar{I} \sim_S PTCI\rangle$$

Obviously it cannot be the case that both of the arguments' claims are true. In this sense these arguments attack, or rebut, each other. We capture this kind of conflict with the following definition.

Definition 4. *If $A = \langle X_A, \lambda_A, \epsilon_A \rangle$ and $B = \langle X_B, \lambda_B, \epsilon_B \rangle$ are two arguments where $\epsilon_A = \tau_1 \circ_I \tau_2$, then we say that A and B attack, conflict with or rebut each other whenever:*

1. *$\epsilon_A = \tau_1 >_I \tau_2$, and $\epsilon_B \in \{\tau_1 \sim_I \tau_2, \tau_2 \sim_I \tau_1, \tau_1 <_I \tau_2, \tau_2 >_I \tau_1\}$.*
2. *$\epsilon_A = \tau_1 \sim_I \tau_2$, and $\epsilon_B \in \{\tau_1 >_I \tau_2, \tau_2 <_I \tau_1, \tau_1 <_I \tau_2, \tau_2 >_I \tau_1\}$.*
3. *$\epsilon_A = \tau_1 <_I \tau_2$, and $\epsilon_B \in \{\tau_1 \sim_I \tau_2, \tau_2 \sim_I \tau_1, \tau_1 >_I \tau_2, \tau_2 <_I \tau_1\}$.*

Note that this definition is symmetric, i.e., if A attacks B then B attacks A.

We will organise the arguments into a graph that we will use as an argumentation framework in the sense of [3]. To do this, we first define an attack relation on arguments in the obvious way: $\mathcal{R}(A, B)$ is true iff A attacks B, for arguments $A, B \in \text{args}(\tau_1, \tau_2, I)$. It is easy to see that the graph induced is tripartite, and its independent sets are $\text{args}(\tau_1 >_I \tau_2), \text{args}(\tau_1 \sim_I \tau_2), \text{args}(\tau_1 <_I \tau_2)$. In our example, this graph would be as follows.

$$\langle \{\gamma_{\text{Armstrong}}\}, \lambda_s, PTC\bar{I} <_S PTCI\rangle$$
$$\Big\downarrow \Big\uparrow$$
$$\langle \{\gamma_{\text{Markman}}\}, \lambda_n, PTC\bar{I} \sim_S PTCI\rangle$$

Since the argument graph is by definition symmetric, it would be beneficial to allow breaking the symmetry with user-defined preferences. We do this by defining preference rules.

Definition 5. *A **preference rule** π is a set of conditions on an ordered pair of conflicting arguments A, B. When the conditions are satisfied, A is said to be preferred to B according to π, and we write $\pi(A, B) = \top$. Otherwise, we say that A is not preferred to B and we write $\pi(A, B) = \bot$.*

We will use π, possibly with subscripts, to denote a preference rule, and Π for the set of preference rules chosen by the user.

Example 5. A preference rule that considers statistically significant results as more authoritative than non significant ones can be seen below.

$$\pi_s(\langle \{\gamma_a\}, \lambda_a, \tau_1 \circ_I \tau_2 \rangle, \langle \{\gamma_b\}, \lambda_b, \tau_1 \diamond_I \tau_2 \rangle) = \top \text{ iff } \begin{cases} \mathsf{sig}(\gamma_a) = \top \\ \text{and } \mathsf{sig}(\gamma_b) = \bot \end{cases}$$

As mentioned previously, $\circ$ and $\diamond$ are meta-variables for the symbols $<, >, \sim$, in other words $\circ$ and $\diamond$ stand for $<, >$ or $\sim$. Here, since by definition the arguments are conflicting, we do not need to specify the values of these meta-variables.

Preference rules are not required to be infallible in any sense. Indeed, π_s embodies one of the aspects of *publication bias*, where by preferring significant results to non-significant ones, one may miss evidence that support the claim that the significant results are a chance occurrence.

We use the preference rules chosen by the user in breaking the symmetry present in $\mathcal{R}$, and capture the new subrelation $\mathcal{R}_\Pi$ of $\mathcal{R}$ as follows.

Definition 6. *For any pair of arguments $A, B \in \mathcal{A}$, $\mathcal{R}_\Pi(A, B)$ is true whenever*

- *$\mathcal{R}(A, B)$ is true and,*
- *if there is a preference rule $\pi \in \Pi$ such that $\pi(B, A) = \top$ then there is also a preference rule $\pi' \in \Pi$ such that $\pi'(A, B) = \top$.*

The motivation here is that if A and B attack each other and A is preferred to B then B's attack on A is cancelled. However, this wording leads to problems when A is preferred to B according to a rule π_1 and B is preferred to A according to π_2. In this case, cancelling both attacks will give the misleading impression that A and B are consistent together. For this reason we give the above, more complicated definition, which only cancels an attack if exactly one argument is preferred to the other.

Now we can put together these components by defining an abstract argumentation framework in the sense of Dung's work [3].

Definition 7. *Given a pair of treatments τ_1, τ_2 and an outcome indicator I, we define the argumentation framework $AF_{\tau_1, \tau_2, I}$ as a pair $\langle \mathcal{A}, \mathcal{R}_\Pi \rangle$ where $\mathcal{A} = \mathrm{args}(\tau_1, \tau_2, I)$.*

Dung [3] defines a notion of admissibility: an admissible set of arguments $\mathcal{S}$ is one which contains no argument that attacks another argument in $\mathcal{S}$ and for any argument $A_1 \in \mathcal{S}$ that is attacked by argument $A_2 \notin \mathcal{S}$, there is another argument $A_3 \in \mathcal{S}$ that attacks A_2. Admissible sets of arguments that are maximal with respect to set inclusion are called *preferred*.

5   Case Study

In this section, we expand our running example and look at how a user might navigate the information contained in the Ovarian Cancer Study. We will see how statistical methods for aggregating results (*meta-analysis*) can be seen as types of arguments that may conflict or agree with other arguments.

In general, a meta-analysis consists of taking a weighted average of a set of results, e.g., a weighted average of hazard ratios. The way the weights are computed depends on statistical assumptions relating to the clinical trials involved. The two main methods, corresponding to different kinds of assumptions, for meta-analyses are the *fixed effects model* and the *random effects model*. Here, we will not look at the details of how each method is defined, but assume that one has been chosen, based on the trial results available. We will use it as a function MA that takes a set of results as argument and returns a *result* that is to be thought of as a virtual trial result. For instance, $\mathsf{hr}(\mathsf{MA}(X))$ represents the hazard ratio of the result produced by a meta-analysis on the set of results X. The inference rule λ_{MA} will then be defined as follows.

λ_{MA}:   Let X be a set of results such that for all $\gamma \in X$, $\mathsf{tr}_1(\gamma) = \tau_1$, $\mathsf{tr}_2(\gamma) = \tau_2$ and $\mathsf{ind}(\gamma) = I$ for given treatments τ_1, τ_2 and outcome indicator I.
　　If $\mathsf{sig}(\mathsf{MA}(X)) = \top$ then:
　　• if $\mathsf{hr}(\mathsf{MA}(X)) < 1$ then $\tau_1 >_I \tau_2$,
　　• if $\mathsf{hr}(\mathsf{MA}(X)) > 1$ then $\tau_1 <_I \tau_2$.
　　Else, if $\mathsf{sig}(\mathsf{MA}(X)) = \bot$ then $\tau_1 \sim_I \tau_2$.

To avoid bias, a meta-analysis should be performed on *all* eligible results, and that is why the definition of λ_{MA} puts such conditions on X.

Returning to the example, below are listed all the trial results reported in the Ovarian Cancer Study that compare the treatments $T_1 = \bar{P}\bar{T}C\bar{I}$ with $T_2 = \bar{P}\bar{T}\bar{C}\bar{I}$. We will denote this set of results as Γ'.

Result	hr ? 1	sig
γTrope	$<$	$\top$
γBarlow	$<$	$\bot$
γBruckner	$<$	$\bot$
γAabo	$<$	$\bot$
γDelgado	$<$	$\bot$
γGronroos	$<$	$\bot$

Result	hr ? 1	sig
γAdams	$<$	$\bot$
γCarmo-Pereira	$>$	$\bot$
γScott	$<$	$\bot$
γPark	$<$	$\bot$
γOmura	$<$	$\bot$
γYoung	$<$	$\bot$

Kyrgiou et al compute the result of the meta-analysis on these trials, and it turns out that $\mathsf{hr}(\mathsf{MA}(\Gamma')) = 0.83 < 1$ and $\mathsf{sig}(\mathsf{MA}(\Gamma')) = \top$.

Using the inference rules $\lambda_s, \lambda_n, \lambda_{MA}$ on these results we get:

$$\mathcal{A}_1 = \mathsf{args}(T_1 >_S T_2) = \left\{ \begin{array}{l} \langle\{\gamma_{\text{Trope}}\}, \lambda_s, T_1 >_S T_2\rangle, \\ \langle\Gamma', \lambda_{MA}, T_1 >_S T_2\rangle \end{array} \right\}$$

$$\mathcal{A}_2 = \mathsf{args}(T_1 \sim_S T_2) = \left\{ \begin{array}{l} \langle\{\gamma_{\text{Barlow}}\}, \lambda_n, T_1 \sim_S T_2\rangle, \\ \vdots \\ \langle\{\gamma_{\text{Young}}\}, \lambda_n, T_1 \sim_S T_2\rangle \end{array} \right\}$$

It should be clear that for every $A_1 \in \mathcal{A}_1$ and $A_2 \in \mathcal{A}_2$, A_1 attacks A_2 and vice versa. We now look at using preference rules with these sets.

Observe that $\mathrm{args}(T_1, T_2, S) = \mathcal{A}_1 \cup \mathcal{A}_2$ and that in the argumentation framework $\langle \mathcal{A}_1 \cup \mathcal{A}_2, \mathcal{R} \rangle$ there are exactly two preferred sets, namely $\mathcal{A}_1$ and $\mathcal{A}_2$. This indicates that we do not have enough information and/or expressed preferences to distinguish between the two possibilities, i.e., $T_1 >_S T_2$ or $T_1 \sim_S T_2$.

Suppose the user applies preference rule π_s, meaning that statistically significant results are preferred to non-significant ones. Then, using $\Pi = \{\pi_s\}$, in the framework $AF_{T_1, T_2, S}$ there remains only one maximal admissible set, $\mathcal{A}_1$, since it contains an argument based on a statistically significant result, γ_{Trope}.

Alternatively, suppose a new preference rule is used that prefers the result of a meta-analysis to a result that is included in the meta-analysis. In other words:

$$\pi_{MA}(\langle X, \lambda_{MA}, \tau_1 \circ_I \tau_2 \rangle, \langle \{\gamma\}, \lambda_x, \tau_1 \diamond_I \tau_2 \rangle) = \top \text{ iff } \lambda_x \in \{\lambda_s, \lambda_n\}$$

Here, λ_x is allowed to be any rule that interprets a single result used in the meta-analysis. Using $\Pi = \{\pi_{MA}\}$, we obtain exactly one preferred set, $\mathcal{A}_1$.

Another emerging method for aggregating trial results is *network meta-analysis*, which is a kind of meta-analysis that uses indirect treatment comparisons as well as direct ones. This means that when there is a trial result $\gamma_{a,b}$ comparing τ_a and τ_b, and another result $\gamma_{b,c}$ comparing τ_b to τ_c, an estimate of the comparison of τ_a and τ_c can be computed, and $\gamma_{a,b}, \gamma_{b,c}$ can be thought of as a path from τ_a to τ_c. Network meta-analyses use such estimates together with direct comparisons (e.g., of τ_a and τ_c) to provide a weighted average of higher inferential power. Without going into details, we use a function $\mathrm{NMA}(\tau_a, \tau_b, X)$, where τ_a and τ_b are the treatments to compare, and X is the set of all results that belong in paths between τ_a and τ_b. Then, we will use an inference rule λ_{NMA} that resembles λ_{MA} in all aspects apart from using a network meta-analysis. We note that acceptance of network meta-analyses as an analytical method is not unanimous and this reinforces the utility of a framework where preferences are user-expressible.

The addition of λ_{NMA} to the set of rules $\Lambda = \{\lambda_s, \lambda_n, \lambda_{MA}\}$ allows for another argument to be generated in our example. The result of the network meta-analysis for T_1 and T_2 was computed in the Ovarian Cancer Study, and it agrees with the standard meta-analysis: $\mathrm{hr}(\mathrm{NMA}(\Gamma'')) = 0.87 < 1$ and $\mathrm{sig}(\mathrm{NMA}(\Gamma'')) = \top$. Note that here we use Γ'' to denote all the results that lie on paths between T_1 and T_2; clearly $\Gamma' \subseteq \Gamma''$. Now, the set $\mathcal{A}_1$ is as follows.

$$\mathcal{A}_1 = \mathrm{args}(T_1 >_S T_2) = \left\{ \begin{array}{l} \langle \{\gamma_{\mathrm{Trope}}\}, \lambda_s, T_1 >_S T_2 \rangle, \\ \langle \Gamma', \lambda_{MA}, T_1 >_S T_2 \rangle, \\ \langle \Gamma'', \lambda_{NMA}, T_1 >_S T_2 \rangle \end{array} \right\}$$

Similarly to the case of the meta-analysis preference rule, one is needed for network meta-analyses too.

$$\pi_{NMA}(\langle X, \lambda_{NMA}, \tau_1 \circ_I \tau_2 \rangle, \langle \{\gamma\}, \lambda_x, \tau_1 \diamond_I \tau_2 \rangle) = \top \text{ iff } \lambda_x \in \{\lambda_s, \lambda_n\}$$

Using $\Pi = \{\pi_{MA}, \pi_{NMA}\}$ yields as preferred the set $\mathcal{A}_1$ once again.

6   Discussion and Conclusions

We have presented a framework for argumentation on treatment efficacy. Its major components are: (1) A language for encoding results from clinical trials, which we believe has many other potential uses such as intelligent querying. (2) A definition delineating the ways arguments on treatment efficacy comparisons can be produced from trial results, along with a definition of a notion of conflict. (3) A definition of a preference rule that allows the potential resolution of conflict depending on the characteristics of the conflicting arguments. Using these components along with standard argumentation tools, users can describe their preferences and analyse the available evidence in terms of agreement and conflict. Finally we have presented the potential use of this framework in a case study on ovarian cancer.

Little work exists that aims to address the problem in focus here. Medical informatics and bioinformatics research does not address the reasoning aspects inherent in the analysis of evidence of primary nature, especially from clinical trials. Previous interesting work ([4,5] and others) exists that uses argumentation as a tool in medical decision support, but as such, assumes the existence of a hand-crafted set of facts around treatment efficacy. Work that is concerned with the capture of a wide spectrum of data about clinical trials exists, [6], and would potentially provide a useful basis for the continuation of our work.

The avenues for further work are several. While not discussed in this paper, the systematic selection and filtering of the trial data used is very important, and requires the formalisation and encoding of many more kinds of meta-data about trials, which we hope to investigate in the future. Such meta-data would include, e.g., the year a trial started (or ended), the location, the healthcare setting, the sample/power size and potentially many other attributes. Prior work exists in this area (see, e.g., [7]) and it would be interesting to see whether argumentation can help in this process as well.

A deeper case study involving clinicians as users would be beneficial in making more concrete the requirements and preferences of such users. Also, liaising with researchers on automated information extraction would provide means to constructing the repositories of trial results that would add significant value to this research. Finally, the construction of a user-friendly system will be beneficial to the evalution and adoption of the framework presented here.

References

1. Kyrgiou, M., Salanti, G., Pavlidis, N., Paraskevaidis, E., Ioannidis, J.: Survival Benefits With Diverse Chemotherapy Regimens for Ovarian Cancer: Meta-analysis of Multiple Treatments. Journal of the National Cancer Institute 98(22), 1655–1663 (2006)
2. Baader, F., Calvanese, D., McGuinness, D., Nardi, D., Patel-Schneider, P. (eds.): The description logic handbook: theory, implementation, and applications. Cambridge University Press, New York (2003)

3. Dung, P.: On the acceptability of arguments and its fundamental role in nonmonotonic reasoning, logic programming and n-person games. Artificial Intelligence 77(2), 321–357 (1995)
4. Patkar, V., Hurt, C., Steele, R., Love, S., Purushotham, A., Williams, M., Thomson, R., Fox, J.: Evidence-based guidelines and decision support services: a discussion and evaluation in triple assessment of suspected breast cancer. British Journal of Cancer 95(11), 1490–1496 (2006)
5. Tolchinsky, P., Cortés, U., Modgil, S., Caballero, F., López-Navidad, A.: Increasing human-organ transplant availability: Argumentation-based agent deliberation. IEEE Intelligent Systems 21(6), 30–37 (2006)
6. Sim, I., Owens, D., Lavori, P., Rennels, G.: Electronic trial banks: A complementary method for reporting randomized trials. Medical Decision Making 20(4), 440–450 (2000)
7. Wyatt, J., Altman, D., Heathfield, H., Pantin, C.: Development of design-a-trial, a knowledge-based critiquing system for authors of clinical trial protocols. Computer Methods and Programs in Biomedicine 43(3-4), 283–291 (1994)

A Knowledge-Management Architecture to Integrate and to Share Medical and Clinical Data, Information, and Knowledge

David Riaño

Research Group on Artificial Intelligence (Banzai), Rovira i Virgili University,
Av. Països Catalans 26, 43007 Tarragona, Spain
david.riano@urv.net

Abstract. Data, information, and knowledge in medicine is varied, changing, interrelated, and for diverse purposes. Medical and Clinical care depends on the correct and efficient combined application of these elements to concrete health care situations as prophylactics, screening, diagnosis, therapy, and prognosis. In this paper, we propose a Knowledge Management Architecture (KMA) to allow the integration of medical and clinical data, information and knowledge in a consistent and incremental way. The components of KMA are described and the already implemented parts are provided with references to papers where they are explained in more detail. For the first time, we present the conceptual integration of the isolated works performed in the research group of artificial intelligence of the Rovira i Virgili University and in collaboration with the Clinical Hospital of Barcelona, and the SAGESSA health care organization.

1 Introduction

Although Knowledge Management (KM) is a concept of business administration, it brings interesting ideas that can be adapted to other domains as the one of health care. KM is a discipline whose main aim is to create, to share, and to apply knowledge in organizations. In this setting, a distinction is made between the concepts of data, information, and knowledge [1]. Data is any value without a context or interpretation (e.g., figure 140). Information refers to context-sensitive and meaningful values that are equivalent to interpreted data (e.g., a patient's blood pressure of 140 mmHg). Finally, knowledge is a generalization that can be applied in several contexts (e.g., patients with blood pressure above 140 mmHg are hypertensive).

Another interesting idea of KM is the classification of knowledge in know-what know-how, and know-why knowledge. Know-what (or declarative) knowledge provides answers to what questions, as for example knowing that *Formoterol*, and *Salmeterol* are the long-acting β_2-agonists recommended for the treatment of Chronic Obstructive Pulmonary Disease (COPD) [2]. Know-how (or procedural) knowledge is useful to answer how questions as for example knowing that when inhaled bronchodilators are not available, regular treatment with slow-release *theophylline* should be considered [2]. Know-why (or explanatory) knowledge is neglected in many application domains, but it is of compelled use in health care, where decisions must be always supported by the greatest evidence possible. This knowledge is designed to

D. Riaño et al. (Eds.): KR4HC 2009, LNAI 5943, pp. 180–194, 2010.

answer why questions as for example the evidence degree and the publications giving support to replacing a short-acting by a long-acting bronchodilator applied to a COPD patient whose *dyspnea* during daily activities is not relieved.

All these concepts (i.e., data, information and know-what, know-how, and know-why knowledge) influence medical and clinical procedures as prophylactics, screening, diagnosis, therapeutics, and prognosis; they provide different perspectives of these procedures, and each one of them complements the other ones.

Medical data and information are based on the standard codification of health care concepts as for example ICD10 [3] for diseases, ATC [4] for drugs, or UMLS [5] for medical concepts in general. These information units are combined to define complex medical and clinical descriptions as for example patient states, medical conditions, or therapies. At the same time these descriptions can be part of more complex information structures as Episodes of Care (EOCs). An EOC refers to the professional management of a health problem from the first encounter with a health care provider through the completion of the last encounter related to that problem [6]. From a structural point of view, chronic comorbid patients define the most complex EOCs in health information systems. Electronic Health Records (EHRs) are EOC-based health information systems that store a longitudinal description of all the medical and clinical information generated for patients visited in a health care center. CEN TC/251 and HL7 RIM are two of the most outstanding EOC-based standards for EHRs.

With regard to medical knowledge, Clinical Practice Guidelines (CPGs) are the health care instruments to gather all the knowledge available to assist practitioners to decide on the appropriate medical and clinical measures in specific circumstances [7]. Unfortunately, apart of the evident pros of CPGs, many have argued that CPGs have some drawbacks. Here, we discus the four of them that affect our work:

1. CPGs are disease-oriented: Gathering all the available evidence about a particular disease has a rather relevant academic interest. However, in daily medical practice the average patient is comorbid and requires simultaneous application of several CPGs by means of the integrated action of several health care professionals.

   On the one hand, this causes that modern health care is defined as patient-oriented (instead of disease-oriented) and consequently the applicability of individual disease-oriented CPGs is reduced. On the other hand, the simultaneous application of several CPGs on the same patient by different professionals, and the lack of health care accepted mechanisms to watch over correct global medical procedures describe a scenario that facilitates medical errors to happen.

2. CPGs are difficult to create and to keep updated: Developing a CPG is an endless task that evolves as new knowledge and evidences appear in relation to the disease the CPG is about (e.g., patient signs and symptoms whose relation with the disease was not observed in previous versions of the CPG, new technologies that allow alternative health care procedures in the CPG, or new pharmacological treatments, contraindications, or ways of application that need to be included in the CPG). This new medical knowledge carries a delay in the modification of CPGs that is caused by clinical trials and legal issues in general but also by the time that the boards of experts modifying the CPGs require to meet and to publish the new updated versions.

3. CPGs are difficult to share: Publishing a good CPG is not equivalent to having a good dissemination of its contents. Public repositories of CPGs as [8], [9], and [10]

are valuable contributors to CPG sharing, but once again this academic approach presents two evident inconveniences in daily medical practice: one is related to the lack of time that physicians have to keep up to date on the CPGs published and also to read them. The second inconvenience is that physicians perceive CPGs as complete views of diseases whereas their professional activities have to do with concrete situations in which only small portions of CPGs are eventually useful. This apparent contradiction between the global approach of CPGs, and the concrete needs of knowledge in daily medical practice is also acting against CPG sharing.

4. CPGs are difficult to put into practice: Related to the previous difficulty, putting CPGs into daily practice becomes a complex task not only because the conceptual distance between the general approach of CPGs to a concrete disease and the concrete approach of physicians to specific patient conditions, but also because the high cost of developing intelligent automated means to help physicians to make decisions according to the statements in a CPG.

In the first case, protocols and care pathways [11] have appeared to bridge the gap between CPGs and medical practice. In the second case, medical informatics has played an important role with the implementation of systems to develop Computer-Interpretable Guidelines (CIG) as for example, Asbru, Proforma, or GLIF3 (see [12] for an overview and comparison). The engineering approach of these systems has been relaxed in other systems as SDA* [13] in order to promote an approach which is closer to medical structures as Clinical Algorithms [14] and to reduce the distance between physicians and formal knowledge representations.

Difficulties 2, 3, and 4 are directly related to the three main aims of KM, which are to create, to share, and to apply knowledge; whereas difficulty 1 is related to both knowledge sharing and knowledge application. Therefore, a KM architecture (KMA) adapted to the medical domain seems to be a promising approach to solve some of the most important drawbacks of CPGs. Moreover, such KMA should be able to incorporate EHR data and information describing EOCs, according to standards.

In this paper, after discussing some related previous works in section 2, a new KM architecture to integrate and to share medical and clinical data, information and knowledge for chronic comorbid patients is presented in section 3, their components introduced and the way they work together explained. Section 4 contains a discussion on the closed and open issues of the process of implementing this health care KMA as a KM Internet platform.

2   Related Work

Previous works on KMAs address the problem of KM with different approaches that range from specific to general purpose. Specific purpose KMAs (also known as vertical KMAs) are based on knowledge processes which are *ad hoc* solutions in a particular domain (e.g., medicine) while general purpose KMAs (also known as horizontal KMAs) define domain-independent models that can be used in different concrete domains. Here, instead of proposing another new specific purpose KMA, our approach was to analyze existing alternative general purpose KMAs, select their best ideas to solve the four medical KM problems discussed in the previous section (i.e., patient-oriented easy creation, updating, sharing, and application of medical knowledge), and then propose a new general purpose KMA adapted to the sorts of medical data, information, and knowledge that we want to manage.

The most common types of general purpose KMAs are either knowledge engineering (e.g., ontology-oriented), incremental, or layered. Knowledge engineering type defines a formalism to represent knowledge (e.g., ontologies) and provides some knowledge processes that are combined in a knowledge life cycle whose purpose is to maintain a knowledge base [15, 16]. Incremental KMAs like [1] define stratified architectures where data is surrounded by information and information by knowledge. Intermediate levels are introduced to link information to data and knowledge to information. KM is implemented on top of knowledge in the outer level of the KMA. Finally, layered KMAs like [15, 17] define abstraction layers containing different knowledge structures and implementing services that are offered to both the KMA final user and also to the immediate upper levels. Typical layers are knowledge source managers, Information Technology infrastructures, document and content managers, organizational taxonomies, general KM services, and knowledge portals.

The knowledge orientation of knowledge engineering KMAs and the high dependence between data, information, and knowledge in incremental KMAs make them less suitable to solve the problems introduced in the previous section than layered KMA where multiple data, information and knowledge from different sources can be progressively incorporated to and updated in the KMA for a later automatic integration and final patient-oriented exploitation and sharing.

As far as health care data, information and knowledge are concerned, several computer-based representation languages have been used in the past. Our choice took into consideration their representation power and capability (they have to deal with complex health care issues), their usability and proximity to the KMA final users (they have to be friendly and easily operated by health care professionals), and their commitment to the Internet (the KM platform implementing the KMA will allow data and knowledge management through the Web).

With these premises, we base our data representation in the principles of CEN TC/251 and HL7 RIM, our know-what representation in OWL DL, our know-how representation in the SDA* language [13] that has been widely applied in real health care settings in the K4CARE project (www.k4care.net), and our know-why representation in XML in order to semantic labeling textual CPGs so that CPG text can be linked to OWL DL classes and SDA* elements. This approach was inspired in [20], but adapted to the sort of information and knowledge structures in our KMA.

3 A Knowledge Management Architecture to Integrate and to Share Medical and Clinical Data, Information, and Knowledge

Medical and clinical activities as prophylactics, screening, diagnosis, therapeutics, or prognosis use knowledge and generate data and information that are stored in EHRs. In Artificial Intelligence, information can be computed to produce knowledge by means of machine learning algorithms. Afterwards and once this knowledge has been validated, it can be used to check whether new arriving information is correct or not by means of automatic reasoning and also to support medical decision making.

In Fig. 1, we propose a multilayered KMA that incorporates medical data, information, and knowledge (see the content management layer) in order to provide a framework to integrate and to share all these elements (knowledge map layer), but also to implement interesting services as healthcare knowledge discovery, explanation seeking, collaboration, personalization or knowledge sharing (service layer).

The architecture also provides a functional interface or knowledge gateway that can be used by external health care systems as e-learning and Randomized Clinical Trial (RCT) tools, web-information systems, consulting systems, quality checking systems, medical research systems, networking systems, etc. (application layer).

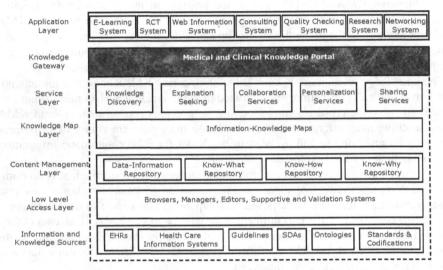

Fig. 1. Multilayered Knowledge Management Architecture

The basic components of the architecture are information and knowledge sources (information and knowledge layer) that are accessible through a set of tools contained in the low-level access layer. This conceptual KMA has been implemented in a KM Internet platform whose users can import, browse, and edit the data and information contained either in the EHRs or Information Systems of health care centers. These heterogeneous sets of information must be accompanied by an explicit representation of their data structures in order to allow the system to incorporate them into a common homogenous representation in the content management layer. This is only possible if they comply with the EOC data model described in section 3.1.

Users can also incorporate, browse and edit know-what knowledge as ontologies, standards and codifications, know-how knowledge as SDA* structures, and know-why knowledge as textual CPGs.

When either new information or knowledge is incorporated to the information and knowledge sources layer an automated process converts it to a homogeneous representation in the content management layer. Then an analysis process is started that searches for relationships between the different parts of this new information (or knowledge) and the other information and knowledge available in that layer. After this complex analysis, the new information (or knowledge) is ready to be included in the content management layer, and the relationships found in the knowledge map layer as it is explained in section 3.5. For example, if data about a new COPD patient is incorporated and the data term *inhaler-training* is not found in the Know-What Repository (i.e., the KMA doesn't know what *inhaler-training* means), the position of the term in the data identifies it as a clinical procedure and it is incorporated as such in the Know-What Repository. The condition and the current treatment of that patient are also searched in both the know-how and know-why repositories and, if they are found, linked as new instances of the corresponding SDA* and CPG sections.

The post-processed data and knowledge elements and their relationships define an integrated data and knowledge base that can be exploited by the tools implemented in the service layer.

In the next sections, the most important parts of this architecture are introduced: the four components of the content management layer in sections 3.1 to 3.4, the knowledge map layer in section 3.5, and two important health care services of the service layer (knowledge discovery and personalization) in section 3.6.

3.1 The EOC Data Model

The EOC Data Model defines the minimal data structure to store an Episode of Care (EOC). An EOC contains a patient identifier and a list of encounters between the patient and any health care professional participating in the EOC.

Each encounter has a different identifier, a date, and a description of the patient condition (i.e., patient state) in terms of a list of descriptors (or state terms). For co-morbid patients, during the same encounter several diseases must be considered. In order to maintain the maximum information possible, the medical actions during the encounter are related to the decision that caused the health care professional to choose these concrete actions. For example, prescribing *formoterol* because of a patient ob-served dyspnea, and ordering a blood analysis because the patient has fever.

The EOC Data Model expressed as a XML schema defines the basic structure of all the data stored in the Data-Information Repository as XML files. XML allows standard exchange of data through the Web, unlike other languages as the Archetype Definition Language (ADL) or the relational data model. More details about the EOC Data Model and the XML Schema can be found in [18].

3.2 The Know-What Ontologies

Knowledge in the Know-What Repository is represented as OWL DL ontologies in order to make all the know-what knowledge used and produced in this KMA usable in Semantic Web applications. These ontologies define the medical vocabulary that is allowed in the KMA data, the information, and the knowledge, but also to formalize the relationships between health care terms. When imported, new ontologies must be OWL DL compliant and their main classes always the same (see Fig. 2). However, additional subclasses and properties are possible to extend the vocabulary of the KMA or to define terms in specific medical settings or institutions.

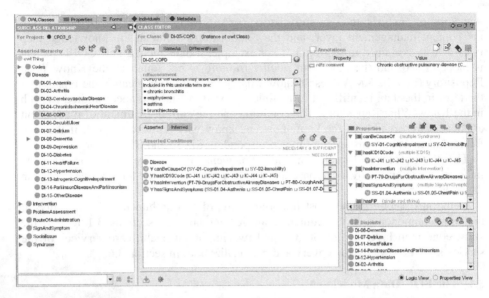

Fig. 2. Example of Ontology in the Know-What Repository of the KM architecture

The main classes are ProblemAssessment (or set of tests to confirm or to find out a medical parameter; for example *spirometry* to calculate the ratio FEV_1/FVC^1 for COPD patients), SignAndSymptom (or set of available signs and symptoms; for example *Asthenia* or *Chest Pain*), Disease (among which the ICD10 codes [3] can be found), and Intervention (among which there are the ATC codes [4]).

The above classes are related with the inverse properties *assesses* (e.g., spirometry assesses FEV_1/FVC), *isSignOrSymptomOf* (e.g., lowFEV1/FVC isSignOrSymptomOf COPD), and *hasIntervention* (e.g., COPD hasIntervention "Drugs for Obstructive Airway Diseases" with ATC code R03).

3.3 The Know-How SDA* Knowledge Model

In this work, procedural knowledge in the Know-How Repository is represented with structures described by the SDA* Knowledge Model [13]. This model represents medical and clinical procedures as a combination of states (S), decisions (D), and actions (A). A state is a set of state terms and it represents all the patients that satisfy all the terms included in the state. A decision is a choice point in which alternative procedures can be followed depending on the decision terms attached to each alternative branch. For example in Fig. 3, the upper diamond indicates alternative treatments depending on whether the patient is *stable* or not. Finally, an action contains a list of interventions (or action terms) for the patient.

---

[1] FEV1/FVC = Forced Expiratory Volume in one second/Forced Vital Capacity. In the ontology, two values are allowed: low FEV1/FVC (<0.70) and normal FEV1/FVC (≥0.70).

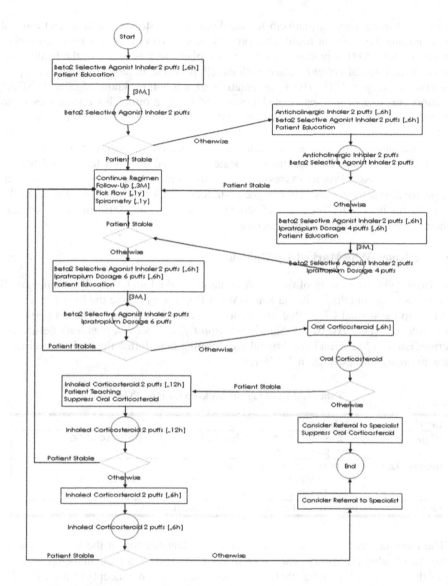

Fig. 3. SDA* structure describing a pharmacological treatment of COPD

States, decisions, and actions are connected describing SDA* structures as the one depicted in Fig. 3 for the pharmacological treatment of COPD. These structures are stored in the Know-How Repository in XML notation. In the K4CARE project[Error! Bookmark not defined.], several European institutions have used the SDA* structures to represent clinical algorithms for fifteen different chronic diseases (i.e., anemia, arthritis, chronic heart failure, cognitive impairment, COPD, coronary heart disease, delirium, dementia, depression, diabetes, hypertension, immobility, Parkinson's, post-stroke, and pressure ulcer). These SDA* structures represent intervention plans that were incorporated in an Internet platform to coordinate and distribute medical and clinical actions among the different health care professionals involved in the care of comorbid patients. In July 2009, a prototype of the whole platform was tested in the health care system of the town of Pollenza, Italy, with excellent results in terms of health care professional satisfaction [21]. The representation of know-how knowledge with SDA* structures was one of the appreciated features of the platform. This encourages us to continue using the SDA* language to represent know-how knowledge in the KMA.

State, decision, and action terms are the instances of some of the classes of the know-what ontology introduced in section 3.2. By default, the instances of the classes SignAndSymptom and Disease are both state terms and decision terms, and the instances of the classes ProblemAssessment and Intervention are action terms. Alternative personalized mappings between the instances of a know-what ontology and the sort of terms in a SDA* can be defined by the user of the platform in the content management layer of the KM architecture.

3.4 The Know-Why Marked-Up Guidelines

The Know-Why Repository of the KMA contains marked-up CPGs. The names of all the classes of the ontologies in the Know-What Repository define the lower level tags to mark-up the textual CPGs that are incorporated to the platform. The properties of the ontologies (i.e., *assesses*, *isSignOrSymptomOf*, and *hasIntervention*) define the intermediate level tags, and conditional and looping tags define the higher level tags. See a summary of these tags in Table 1.

Table 1. List of tags to mark-up textual CPGs

Level	XML Tags
Higher	`<if><decision></decision><action><action/></if>` `<loop></loop>`
Intermediate	`<assesses></assesses>` `<isSignOrSymbolOf></isSignOrSymbolOf>` `<hasIntervention></hasIntervention>`
Lower	the name of the classes in the ontologies

The mark-up process starts with the syntactic identification of the instance names of the know-what repository ontologies in the CPG text, and their mark-up with the tag of the instance class. After that, consecutive near tagged concepts in the CPG that share a relationship in the ontology are analyzed in order to detect tags of the sort `assesses`, `isSignOrSymbolOf`, or `hasIntervention`. If these relations exist, the tags are inserted in the CPG. Finally, higher level tags are searched by looking for if-then and loop expressions in the text.

The last two steps of the process are complex tasks that we have not completely solved by automatic natural language processing (NLP) methods [19]. At the moment, the cases that are still not solved with NLP are decided by hand.

See an example of mark-up in Fig. 4 in which concepts as "mild COPD", "dyspnea", and "inhaled bronchodilators" are found in the ontology as Disease, SignAndSymptom and ATC code, respectively. The *isSignOrSymbolOf* relationship

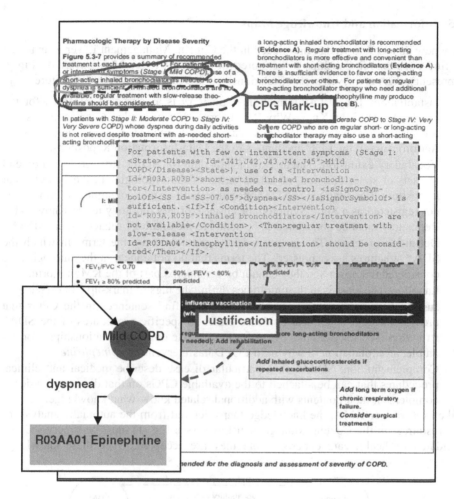

Fig. 4. Page 54 of GOLD CPG Initiative [2] as a know-why resource to justify part of a SDA* structure in the Know-How Repository. Observe that *Epinephrine* (R03AA01) is a kind of short-acting inhaled bronchodilator (R03A).

between "dyspnea" and "mild COPD" is found in the know-what ontology of Fig. 2 and, therefore, the tag <isSignOrSymbolOf> is introduced in the textual CPG. Finally, an if-then expression between "inhaled bronchodilators are not available" (i.e., decision) and "regular treatment with slow-release theophyline" (i.e., action) is detected and marked-up as such.

The incorporation of textual CPGs in the Information and Knowledge Sources layer, their automatic (still semi-automatic) mark-up in the Know-Why Repository, and their final integration with the KMA data, information, know-what, and know-how knowledge through knowledge maps in the Knowledge Map Layer contributes to make CPG update and usage easier.

3.5 Information and Knowledge Maps

Any new data or knowledge introduced in the Content Management Layer is analyzed to find out new relationships to be stored in the Knowledge Map Layer. As Fig. 6 represents, there are seven sorts of relationships that this layer can find and store:

- Instantiation: The Data-Information Repository is a source of instances for the Know-How and the Know-Why Repositories. For example, instances of a state-decision-action sequence in a SDA* can be found in the Condition and (Decision, Action) pairs of the EOC encounters in the Data-Information Repository.
- Vocabulary extension: The arrival of new data (or SDA* structures) can extend the sets of state, decision, or action terms of the ontologies in the Know-What Repository. For example, a SDA* structure with a state containing a term that is not registered as a Sign, Symptom or Disease of any ontology in the Know-What Repository must be considered to be added to the ontology related to that SDA*.
- Definition of terms: The instances in the ontologies are the terms in which the SDA* structures are expressed. The terms that do not appear in the ontologies are considered unknown and they cannot be considered part of the KM platform.
- Labeling: The classes in the ontologies define the tags to mark-up CPGs (see 3.4)
- Justification and Explanation: NLP is used to find sentences in the CPGs that provide medical or clinical evidence either to specific procedures of the SDA* structures (see for example, Fig. 4) or to concrete ontology relationships. For example, an explanation of why mild COPD hasIntervention *epinephrine*.
- Complementation: SDA*s on a particular disease describe medical and clinical procedures that can be attached to the available CPGs on that disease in order to complement their contents with additional related know-what knowledge.

Like in any other KMA, the knowledge maps derived from the automatic analysis of the resources in the Content Management Layer (see Fig. 1) must be supervised and validated by health care experts before they are accepted. Once these knowledge

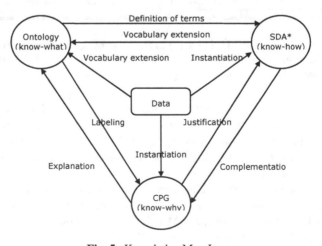

Fig. 5. Knowledge Map Layer

maps are validated they can be used to support the tools that are available in the Service Layer.

3.6 The Service Layer

The Service Layer hosts a set of tools (see Fig. 1) among which knowledge discovery and personalization are the only ones discussed here.

Knowledge Discovery. There are three sorts of knowledge in the KM platform: know-what (i.e., OWL DL ontologies), know-how (i.e., SDA* structures), and know-why (i.e., marked-up CPGs). The information in the Data-Information Repository is analyzed with inductive Machine Learning methodologies in order to derive ontology components (i.e., new terms and relationships) and SDA* structures (i.e., patient states, treatment actions, and their combination through SDA* decisions). An efficient accurate algorithm to induce SDA* structures from the data in the Data-Information Repository is introduced in [22]. It is based on the idea of detecting representative patient states and treatment action blocks in the data and then connecting them with decision trees whose leaves are either states (definition of patient evolutions) or action blocks (definition of patient treatments). The result is a SDA* structure that integrates the long term treatments observed in the data of the Data-Information Repository.

Personalization. In health care, knowledge personalization entails the application of two different technologies. On the one hand, personalization consists on the simplification of the available knowledge models to the particularities of the targeted patient. So, for example, if a patient has *moderate COPD*, all the ontological concepts and treatments related exclusively to severe and very severe COPD should be removed from the knowledge available for that patient. On the other hand, if the patient is co-morbid the knowledge of each one of the patient diseases have to be merged or combined into a single unit representing the knowledge about this patient. In the past, we considered alternative approaches to these sorts of personalization [23], but in our current KM platform, personalization has been implemented in the context of the K4CARE project [24, 25]. There, the health care terms that are not related to a concrete patient are hidden in the know-what ontologies (e.g., diseases that the patient do not have), and in the know-how SDA*s (e.g., states which the patient cannot be in). At the end of the process, this personalization service offers a view of the knowledge in the Content Management Layer which is personalized to each single patient, providing patient-oriented health care, instead of disease-oriented.

4 Discussion

This paper introduces a new KMA aiming at the automatic integration and sharing of medical and clinical data, information, and knowledge. These integrated formal structures are employed to define knowledge maps that are useful to provide health care services as knowledge discovery and personalization. This KMA is progressively implemented as a KM platform that constitutes the general framework of the members of the research group on artificial intelligence at Rovira i Virgili University in Tarragona, Spain. The KMA is arranged in layers, among which the content management layer (i.e., data, information [18], and knowledge models [13]) has been thoroughly tested in past works [21,22,25], the knowledge map layer is partly finished [19], and some tools of the service layer are at different levels of development [19, 22, 25].

The four CPG problems described in the introduction can be addressed with the KMA presented: a *patient-oriented* rather than a disease-oriented medical and clinical knowledge is possible with the KMA in two ways: retrospective and prospective. The data about the EOCs of concrete patients introduced in the data-information repository is linked to know-what, know-how, and know-why knowledge in the Knowledge Map Layer providing all the knowledge available about the what, how, and why of past treatments allowing retrospective analysis. Moreover, the personalization service adapts the medical and clinical knowledge in the Content Management Layer to new patients providing a prospective orientation of knowledge.

The difficulties *to create and to keep medical knowledge updated* can be simplified with the use of the KMA. On the one hand, new CPGs can be not only stored in the know-why repository of the KM Internet Platform, but also automatically integrated with the knowledge available in the Content Management Layer. This computerized process is expected to reduce CPG knowledge publication delays. On the other hand, the data, information and knowledge in the KM platform can also interact with RCT systems in order to prepare, conduct, and analyze RCTs (see Fig. 1). Successful results of RCT can then be easily appended to the Content Management Layer as evidence-based knowledge, and integrated with the rest of available knowledge.

KMA addresses the problem of *knowledge sharing* providing Internet access to medical and clinical knowledge through the KM Platform. The Service Layer also provides tools to approach global health care knowledge to concrete situations in daily medical practice (e.g., explanation and personalization services). Moreover, these services allow KMA users *to put medical and clinical knowledge into practice* with the combined use of computer-executable SDA*s and personalization services.

The current open issues are (1) to conclude the implementation of the Knowledge Map Layer, (2) to extend the know-what repository to contain the interaction between drugs and also the interaction between drugs and patient conditions in order to improve the quality of some of the services provided in the Service Layer (e.g., merging SDA*s [25] to personalize the treatment of comorbid patients), and (3) to complete the evaluation of the benefits of the KMA with respect to the problems explained in section 1 in two concrete settings: the Clinical Hospital of Barcelona and the SAGESSA health care organization in Tarragona; where we are working on the diseases of hypertension, heart failure, COPD, diabetes, and dyslipidemia.

References

1. Müller, H.J., Schappert, A.: The Knowledge Factory – A Generic Knowledge Management Architecture. In: Proceedings of IJCAI Workshop on Knowledge Management and Organisational Memory, Stockholm (1999)
2. Buist, A.S. (chair): Global Initiative for Chronic Obstructive Lung Disease. Global Strategy for the Diagnosis, Management, and Prevention of Chronic Obstructive Pulmonary Disease (2007), http://www.who.int/respiratory/copd/GOLD_WR_06.pdf (Last accessed: October 2009)

3. International Statistical Classification of Diseases and Related Health Problems, 10th Revision. World Health Organization, WHO (2007),
http://apps.who.int/classifications/apps/icd/icd10online/
(Last accessed: October 2009)
4. Anatomical Therapeutic Chemical (ATC), 6th Edition. World Health Organization, WHO (2003), http://www.who.int/classifications/atcddd/en/ (Last accessed: October 2009)
5. Unified Medical Language System (UMLS), US National Library of Medicine (2009), http://www.nlm.nih.gov/research/umls/ (Last accessed: October 2009)
6. Donaldson, M., Yordy, K., Vanselow, N. (eds.): Defining primary care: an interim report. National Academy Press, Washington (1994)
7. Field, M.J., Lohr, K.H. (eds.): Clinical Practice Guidelines: Directions for a New Program, Institute of Medicine. National Academy Press, Washington (1990)
8. US National Guideline Clearinghouse,
http://www.guideline.gov (Last accessed: October 2009)
9. UK eGuidelines Site, http://www.eguidelines.co.uk (Last accessed: October 2009)
10. CMA, http://www.cma.ca/index.cfm/ci_id/54296/la_id/1.htm (Last accessed: October 2009)
11. Campbell, H., Hotchkiss, R., Bradshaw, N., Porteous, M.: Integrated care pathways. BMJ. 316(7125), 133–137 (1998)
12. Peleg, M., Tu, S., Bury, J., Ciccarese, P., Fox, J., Greenes, R., et al.: Comparing Computer Interpretable Guideline Models: A Case-Study Approach. JAMIA 10(1), 52–68 (2003)
13. Riaño, D.: The SDA* Model: A Set Theory Approach. In: Proc. of the 20th IEEE International Symposium on Computer-Based Medical Systems, pp. 563–568 (2007)
14. Society for Medical Decision Making Committee on Standardization of Clinical Algorithms: Proposal for Clinical Algorithm Standards. Med. Decis. Making 12, 149–154 (1992)
15. Lindvall, M., Rus, I., Sinha, S.S.: Software systems support for knowledge management. Journal of Knowledge Management 7(5), 137–150 (2003)
16. Juarez, J.M., Riestra, T., Campos, M., Morales, A., et al.: Medical knowledge management for specific hospital departments. Exp. Sys. App. 36, 12214–12224 (2009)
17. Lawton, G.: Knowledge Management: Ready for Prime Time? IEEE Comp. 34(2), 12–14 (2001)
18. Riaño, D., Bohada, J.A., Real, F., López-Vallverdú, J.A.: The SDA* Data Model. HYGIA Project. Internal Report,
http://banzai-deim.urv.net/~riano/
TIN2006-15453/Documentos/DataModel01.pdf (Last accessed: October 2009)
19. Taboada, M., Meizoso, M., Riaño, D., et al.: From Natural Language Descriptions in Clinical Guidelines to Relationships in an Ontology. In: Riaño, D., ten Teije, A., Miksch, S., Peleg, M. (eds.) Proc. of the AIME 2009 Workshop on Knowledge Representation for Health-Care. LNCS (LNAI), vol. 5943. Springer, Heidelberg (2010)
20. Seyfang, A., Martínez-Salvador, B., Serban, R., Wittenberg, J., Miksch, S., Marcos, M., ten Teije, A., Rosenbrand, K.: Maintaining Formal Models of Living Guidelines Efficiently. In: Bellazzi, R., Abu-Hanna, A., Hunter, J. (eds.) AIME 2007. LNCS (LNAI), vol. 4594, pp. 441–445. Springer, Heidelberg (2007)
21. Tomassini, P., Campana, F., et al.: D12.2 Validation and Assessment of platform – II (2009), http://www.k4care.net (Last accessed: October 2009)

22. Riaño, D., López-Vallverdú, J.A., Tu, S.: Mining Hospital Data to Learn SDA* Clinical Algorithms. In: Riaño, D. (ed.) K4CARE 2007. LNCS (LNAI), vol. 4924, pp. 46–61. Springer, Heidelberg (2008)
23. Abidi, S.R., Abidi, S.S.: Towards the Merging of Multiple Clinical Protocols and Guidelines via Ontology-Driven Modeling. In: Proc. of the ECAI 2008 Workshop Knowledge Management for HealthCare Processes. LNCS (LNAI), vol. 5651, pp. 81–85. Springer, Heidelberg (2009)
24. Real, F., Riaño, D.: An Autonomous Algorithm for Generating and Merging Clinical Algorithms. In: Proc. of the ECAI 2008 Workshop Knowledge Management for HealthCare Processes. LNCS (LNAI), vol. 5626, pp. 13–24. Springer, Heidelberg (2009)
25. Isern, D., Moreno, A., Pedone, G., Sánchez, D., Varga, L.Z.: Home Care Personalization with Individual Intervention Plans. In: Proc. of the ECAI 2008 Workshop Knowledge Management for HealthCare Processes. LNCS (LNAI), vol. 5626, pp. 134–151. Springer, Heidelberg (2009)

Author Index

Abidi, Syed Sibte Raza   88
Acosta, Dionisio   124
Aleksovski, Zharko   50
Alonso, Albert   26, 113
Alonso, Josep Ramon   113

Bottrighi, Alessio   76

Cardillo, Elena   38
Cavallini, Anna   141
Chesani, Federico   76

Daniyal, Ali   88

Eccher, Claudio   14

Ferro, Antonella   14
Fox, John   124

Gatt, Albert   155
Gorogiannis, Nikos   169

Hunter, Anthony   169

Keshtgar, Mo   124

Lozano, Esther   113
Lucas, Peter J.F.   100

Marcheselli, Simona   141
Marcos, Mar   113
Martínez, Diego   26
Martínez-Salvador, Begoña   113
Meizoso, María   26

Mello, Paola   76
Micieli, Giuseppe   141
Miksch, Silvia   14
Milian, Krystyna   50
Montali, Marco   76
Montani, Stefania   76
Moreno, Antonio   1

Panzarasa, Silvia   141
Patkar, Vivek   124, 169
Peleg, Mor   64
Portet, François   155

Quaglini, Silvana   141

Riaño, David   26, 180

Sánchez, David   1
Serafini, Luciano   38
Seyfang, Andreas   14
Stankevich, Sergey   14
Stefanelli, Mario   141
Storari, Sergio   76

Taboada, María   26
Tamilin, Andrei   38
ten Teije, Annette   50
Terenziani, Paolo   76

van der Heijden, Maarten   100
van Harmelen, Frank   50
Vdovjak, Richard   50

Williams, Matthew   169

Author Index